Lecture Notes
in Control and Information Sciences   453

Editors

Professor Dr.-Ing. Manfred Thoma
Institut fuer Regelungstechnik, Universit  nover,
Germany
E-mail: thoma@irt.uni-hannover.de

Professor Dr. Frank Allgöwer
Institute for Systems Theory and Autom
Pfaffenwaldring 9, 70550 Stuttgart, Germany
E-mail: allgower@ist.uni-stuttgart.de

Professor Dr. Manfred Morari
ETH/ETL 129, Physikstr. 3, 8092 Zürich, Switzerland
E-mail: morari@aut.ee.ethz.ch

For further volumes:
www.springer.com/series/642

Francesco Amato · Roberto Ambrosino ·
Marco Ariola · Carlo Cosentino ·
Gianmaria De Tommasi

Finite-Time
Stability
and Control

 Springer

Francesco Amato
Dipartimento di Medicina Sperimentale
  e Clinica
Università Magna Graecia di Catanzaro
Catanzaro, Italy

Roberto Ambrosino
Dipartimento di Ingegneria
Università degli Studi di Napoli Parthenope
Napoli, Italy

Marco Ariola
Dipartimento di Ingegneria
Università degli Studi di Napoli Parthenope
Napoli, Italy

Carlo Cosentino
Dipartimento di Medicina Sperimentale
  e Clinica
Università Magna Graecia di Catanzaro
Catanzaro, Italy

Gianmaria De Tommasi
Dipartimento di Ingegneria Elettrica
  e delle Tecnologie dell'Informazione
Università degli Studi di Napoli Federico II
Napoli, Italy

ISSN 0170-8643   ISSN 1610-7411 (electronic)
Lecture Notes in Control and Information Sciences
ISBN 978-1-4471-5663-5   ISBN 978-1-4471-5664-2 (eBook)
DOI 10.1007/978-1-4471-5664-2
Springer London Heidelberg New York Dordrecht

Library of Congress Control Number: 2013956519

Printed on acid-free paper

Springer is part of Springer Science+Business Media (www.springer.com)

Preface

In the last fifteen years a great effort has been spent by control researchers to investigate the finite-time control problem for dynamical systems. The main objective of this book is that of presenting, in a unified framework, the main results appeared in the literature on this topic, with particular reference to the finite-time stability problems involving linear time-varying systems and hybrid systems whose continuous-time dynamics are linear time-varying.

We essentially consider three main topics: finite-time stability analysis and design for linear time-varying systems (both continuous-time and discrete-time) with ellipsoidal initial and trajectory sets; finite-time stability analysis for linear time-varying systems when the initial and trajectory sets are assumed to be piece-wise quadratic; finite-time stability analysis and control for hybrid systems whose continuous-time dynamics are described by a linear time-varying system. For each main problem, except for the case where piecewise quadratic sets are considered, the results are given in the form of necessary and sufficient conditions involving the feasibility of differential linear matrix inequalities and problems based on differential Lyapunov equations.

For self-containedness purposes, most of the results provided in the book are proven. We have tried to maintain the development of the proofs as simple as possible, without sacrificing the mathematical rigor.

The work is essentially devoted to both researchers in the field of systems and control theory and engineers working in industries who want to apply the methodologies presented in the book to practical control problems. To this regard, as the various results are derived, they are immediately reinforced with real-world examples.

Catanzaro, Napoli
September 2013

Francesco Amato
Carlo Cosentino
Roberto Ambrosino
Marco Ariola
Gianmaria De Tommasi

Contents

Acronyms

Abbreviations

CT-LTI	Continuous-time linear time-invariant
CT-LTV	Continuous-time linear time-varying
DT-LTI	Discrete-time linear time-invariant
DT-LTV	Discrete-time linear time-varying
DLE	Differential (difference) Lyapunov equation
DLMI	Differential (difference) linear matrix inequality
D/DLE	Differential/Difference Lyapunov equation
D/DLMI	Differential/Difference linear matrix inequality
FTS	Finite-time stability
IDLS	Impulsive dynamical linear systems
IO-FTS	Input–output finite-time stability
LFT	Linear fractional transformation
LMI	Linear matrix inequality
LS	Lyapunov stability
PQLF	Piecewise quadratic Lyapunov function
PQD	Piecewise quadratic domain
QFTS	Quadratic finite-time stability
RFTS	Robust finite-time stability
SD-IDLS	State-dependent IDLS
TD-IDLS	Time-dependent IDLS
TD-SLS	Time-dependent switching linear system

Mathematical Symbols

:	such that
$\forall$	for all
$\exists$	there exists
$:=$	equal by definition
$p \Leftrightarrow q$	p is equivalent to q
$p \Rightarrow q$	p implies q

Set Theory

$x \in A$	The element x belongs to the set A
$S_1 \cup S_2$	The union of the sets S_1 and S_2
$S_1 \subseteq S_2$	The set S_1 is a subset of the set S_2
$S_1 \subset S_2$	The set S_1 is a *strict* subset of the set S_2
S/T	the set composed of the points that belong to S and do not belong to T
$\text{conv}(E)$	Convex hull of a set E
$\text{cone}(\{x_1, \ldots, x_p\})$	the convex set (cone) generated by the conic combination of vectors $x_1, \ldots, x_p$

Numerical Sets

$\mathbb{N}$	Nonnegative (positive) integer numbers
$\mathbb{R}$	Field of real numbers
$\mathbb{R}^n$	Set of the n-tuples of real numbers
$\mathbb{R}^{m \times n}$	Real matrices with m rows and n columns

Vector and Matrix Operators

x_i	The ith element of a vector x
a_{ij}	The ijth element of a matrix A
A^{-1}	Inverse of a square matrix A
A^T	Transpose of a matrix A
$\text{diag}(A_1, A_2, \ldots, A_r)$	Block diagonal matrix with $A_1, A_2, \ldots, A_r$ on the diagonal
$A > 0$	A is (symmetric) positive definite
$A \geq 0$	A is (symmetric) positive semidefinite
$A < 0$	A is (symmetric) negative definite
$A \leq 0$	A is (symmetric) negative semidefinite
$A > B$	$A - B$ is (symmetric) positive definite
$A \geq B$	$A - B$ is (symmetric) positive semidefinite
$\text{rank}(A)$	The rank of a matrix A
$A \otimes B$	Kronecker product of matrices A and B

Norms

$\|x\|$	Euclidean norm of the vector $x \in \mathbb{R}^n$ $(= \sqrt{\sum_{i=1}^n x_i^2})$
$\|x\|_\infty$	Infinity norm of the vector $x \in \mathbb{R}^n$ $(= \max\{|x_1|, \ldots, |x_n|\})$
$\|A\|$	Spectral norm of a matrix A (i.e., the maximum singular value of A)

Miscellaneous

$\square$	End of theorems, lemmas, corollaries, and facts
$\triangle$	End of examples
$\diamond$	End of assumptions, definitions, problems, procedures, exercises, remarks, and proofs
wrt	With respect to

Chapter 1
Introduction

1.1 A Brief History of Finite-Time Stability

The concept of finite-time stability (FTS) dates back to the 1950s, when it was introduced in the Russian literature [62, 64, 65]; later, during the 1960s, this concept appeared in the western journals [44, 72, 85]. Roughly speaking, a system is said to be finite-time stable if, given a bound on the initial condition, its state does not exceed a certain threshold during a specified time interval. More precisely, given the system

$$\dot{x}(t) = f(t, x), \quad x(t_0) = x_0, \qquad (1.1)$$

where $x(t) \in \mathbb{R}^n$, we can give the following formal definition [6], which restates the original definition in a way consistent with the notation of this book.

Definition 1.1 (FTS) Given an initial time t_0, a positive scalar T, two sets $\mathcal{X}_0$ and $\mathcal{X}_t$, system (1.1) is said to be finite-time stable wrt $(t_0, T, \mathcal{X}_0, \mathcal{X}_t)$ if

$$x_0 \in \mathcal{X}_0 \Rightarrow x(t) \in \mathcal{X}_t, \quad t \in [t_0, t_0 + T], \qquad (1.2)$$

where, with a slight abuse of notation, $x(\cdot)$ denotes the solution of (1.1) starting from x_0 at time t_0.   ◊

   In the following, we shall mainly consider linear time-varying (LTV) systems; however, in the second part of the book, we shall extend the concept of FTS to hybrid systems. The sets $\mathcal{X}_0$, the *initial set* (or domain), and $\mathcal{X}_t$, the *trajectory set* (or domain), will be either ellipsoids, when quadratic Lyapunov functions are involved, or polytopes (more generally, piecewise quadratic domains (PQDs)), when piecewise quadratic Lyapunov functions (PQLFs) are used.
   Note that the trajectory set is allowed to vary in time. For well-posedness of Definition 1.1, it is required that $\mathcal{X}_0 \subset \mathcal{X}_{t_0}$. However, in principle, it is not required that $\mathcal{X}_0$ is included in $\mathcal{X}_t$ for $t > t_0$.
   In the last fifteen years, FTS and stabilization have been investigated in the context of continuous-time linear systems (e.g., see [4–6, 8, 12, 16, 20, 23, 49, 78])

F. Amato et al., *Finite-Time Stability and Control*,
Lecture Notes in Control and Information Sciences 453,
DOI 10.1007/978-1-4471-5664-2_1, © Springer-Verlag London 2014

and of discrete-time linear systems (see, among others, [2, 9, 14, 18]); in these papers, conditions for analysis and design are generally provided in terms of feasibility problems involving Linear Matrix Inequalities (LMIs) [38] and/or Differential Linear Matrix Inequalities (DLMIs) [77] and Differential Lyapunov Equations (DLEs). More recently, an effort has been spent to extend such results to the context of nonlinear systems (e.g. see [20, 71, 88]) and hybrid systems [15, 27, 28, 41, 82, 88, 89, 91].

1.2 FTS, Lyapunov Stability, and Related Issues

Why is the property expressed by (1.2) called FTS? To answer this question, we recall the classical definition of Lyapunov stability (LS) [55, 63]. System (1.1) is said to be stable in the sense of Lyapunov if, for all $\epsilon > 0$, there exists a positive scalar δ, possibly depending on t_0 and ϵ, such that $\|x_0\| < \delta(\epsilon, t_0)$ implies

$$\|x(t)\| < \epsilon, \quad t \geq t_0.$$

The key points in the above definition are: the system is stable if, once an arbitrary value for ϵ has been fixed, it must be possible to build an inner ball (of radius δ) such that, whenever the initial condition is inside such a ball, the trajectory of the system starting from x_0 does not exit the outer ball (of radius ϵ); moreover, this property holds for all t between t_0 and infinity. Note that LS is a qualitative concept, that is, both the inner and the outer ball are not quantified; therefore, LS can be regarded as a structural property: either a system is stable, or it is not.

Now let us come back to Definition 1.1; even in this case, we have an inner set, namely $\mathcal{X}_0$, and an outer set, $\mathcal{X}_t$, and it is required that whenever the trajectory starts inside the inner set, it does not exit the outer set. From this point of view, Definition 1.1 mimics the one of classical Lyapunov stability, and this justifies the use of the term *stability*; however, an important point is that, differently from classical stability, this is only required over a *finite*, possibly short with respect to steady state, interval of time. Another important point is that FTS is a quantitative concept, since the inner and the outer sets are specified once and for all. Therefore the same system can be finite-time stable for some choice of $\mathcal{X}_0$, $\mathcal{X}_t$ and T and non-finite-time stable for a different choice of these sets and/or parameters.

As a consequence, FTS and LS are independent concepts; indeed, a system can be finite-time stable but not stable in the sense of Lyapunov and vice versa. While LS deals with the behavior of a system within a sufficiently long (in principle, infinite) time interval, FTS is a more practical concept, useful to study the behavior of the system within a finite (possibly short) interval, and therefore it finds application whenever it is desired that the state variables do not exceed a given threshold (for example, to avoid saturations or the excitation of nonlinear dynamics) during the transients.

The FTS dealt with in this book should not be confused with the finite-time stability concept adopted in some other works such as, for instance, [34, 56, 74] for autonomous systems and [57, 73] for nonautonomous systems. In these papers, the authors focus on the stability analysis of nonlinear systems whose trajectories converge to an equilibrium point in finite time and on the characterization of the associated *settling-time*. In this context, *finite-time stability implies LS* together with the convergence of the system trajectories to an asymptotically stable equilibrium state in finite time. In fact, it turns out that this alternative definition of finite-time stability is a *stronger* notion than asymptotic stability; hence, it is unrelated to the one given in [44, 85] and considered in this book. Furthermore, the concept of finite-time stability given in [34, 56, 74] would not be well posed in the context of continuous-time linear systems, since the state of such a system cannot go to zero by using a linear controller.

Since the concept of FTS is, in some way, related to the concept of *reachable sets*, it is important to clarify the main differences between the two ideas. A reachable set is defined as a set of states that a dynamical system attains given some bounded inputs and starting from given initial conditions; this concept was proposed in [52]. In other words, reachable set analysis answers the following question (see also [53]): *Which set of states is reached by a system starting from a given set of initial conditions and subject to bounded inputs?* On the other hand, FTS analysis answers the following question: *Given a bound on the state variables and a set of admissible initial states, does the state remain confined within the prescribed bound when nonzero initial conditions are considered?* Therefore, one main difference between the two concepts is that reachable set theory considers the presence of inputs whereas FTS does not; moreover, in the reachable set analysis, the assumption of system asymptotic stability is necessary (see, for example, [38, Chap. 6] and [50]), while, as discussed above, the FTS analysis condition provided in this book will allow us to deal also with systems that are not asymptotically stable.

1.2.1 FTS: An Input–Output Perspective

In more recent years, the concept of input–output finite-time stability (IO-FTS) has been introduced in the control literature (see [17, 26]). In short, a system is said to be IO-finite-time stable if, given a class of norm bounded input signals defined over a specified time interval, the output of the system does not exceed an assigned threshold during such a time interval.

While FTS is, in a sense, related to classical LS (see Sect. 1.2), the definition of IO-FTS can be framed in the context of bounded input–bounded output stability. We recall that a system is said to be IO $\mathcal{L}_p$-stable [63, Chap. 5] if for any input of class $\mathcal{L}_p$, the system exhibits a corresponding output that belongs to the same class.

The main differences between *classic* IO stability and IO-FTS are that the latter involves signals defined over a finite time interval, does not necessarily require the inputs and outputs to belong to the same class, and that *quantitative* bounds on both inputs and outputs must be specified; therefore, IO stability and IO-FTS are independent concepts. While IO stability deals with the behavior of a system within a sufficiently long (in principle infinite) time interval, IO-FTS is a more practical concept, useful to study the behavior of the system within a finite (possibly short) interval, and therefore it finds application whenever it is desired that the output variables do not exceed a given threshold during the transients, given a certain class of input signals.

The definition of IO-FTS given in [17] is fully consistent with the definition of (state) FTS, given in Sect. 1.1. In [17] two sufficient conditions for IO-FTS were provided for the class of $\mathcal{L}_2$ inputs and the class of $\mathcal{L}_\infty$ inputs, respectively. Both conditions required the solution of a feasibility problem involving DLMIs.

More recently, in [26], it has been shown that, when $\mathcal{L}_2$ inputs are considered, the condition given in [17] is actually also *necessary*. In the same paper, an alternative necessary and sufficient condition requiring that a certain DLE admits a positive definite solution is provided; to prove the last results, a machinery involving the concept of reachability Gramian is used. It is shown that the condition based on the DLE is more effective from a numerical point of view; on the other hand, the DLMI formulation turns out to be useful for design purposes.

As in the FTS case, it should be mentioned that a different concept of IO-FTS has been recently given for nonlinear systems. In particular, the authors of [57] consider systems with a norm bounded input signal over the interval $[0, +\infty]$ and a nonzero initial condition. In this case, the IO-FTS is related to the property of a system to have a norm bounded output that, after a finite time interval of length T, does not depend anymore on the initial state. Hence, the concept of IO-FTS introduced in [17] and the one in [57] are different. Note again that the definition of IO-FTS given in [57] would not be well posed in the context of continuous-time linear systems since the output of such a system cannot go to zero in finite time (when nonimpulsive inputs are considered).

1.3 Book Organization

This book is composed of two parts and nine chapters. Part I deals with linear systems, and Part II with hybrid systems.

Part I starts with Chaps. 2 and 3, where an approach to analysis and design of linear systems based on quadratic Lyapunov functions is considered.

In particular, continuous-time linear time-varying (CT-LTV) systems of the form

$$\dot{x}(t) = A(t)x(t) + B(t)u(t), \quad x(t_0) = x_0, \tag{1.3a}$$

$$y(t) = C(t)x(t) + D(t)u(t), \tag{1.3b}$$

are considered.

For CT-LTV systems, some conditions guaranteeing FTS and finite-time stabilization will be presented. The proposed approach will make use of time-varying quadratic Lyapunov functions and will lead to necessary and sufficient conditions for analysis and synthesis based on feasibility problems involving DLMIs and DLEs.

In Chap. 4 the robustness issues are considered, while a generalization to discrete-time systems of the above results is presented in Chap. 5.

Chapter 6 contains the most recent advancements of FTS theory; indeed, it will be devoted to illustrate techniques based on nonquadratic Lyapunov functions. In particular, we shall focus on PQLFs. The derived sufficient conditions for the FTS analysis will be shown to be less conservative than the ones derived by the approaches based on quadratic Lyapunov functions, when the Initial and Trajectory sets are assumed to be polytopic (or, more generally, piecewise quadratic).

Part II starts with Chap. 7, where the FTS analysis for a special class of hybrid systems, namely impulsive dynamical linear systems (IDLSs) [54], is considered. An IDLS is a CT-LTV system whose state undergoes finite jump discontinuities at discrete instants of time. Chapter 8 deals with the design problem for the class of such systems.

IDLSs have the form

$$\dot{x}(t) = A_c(t)x(t) + B(t)u(t), \quad (t, x(t)) \notin \mathcal{S}, \tag{1.4a}$$

$$x(t_k^+) = A_d(t_k)x(t_k), \quad (t_k, x(t_k)) \in \mathcal{S}, \tag{1.4b}$$

$$y(t) = C(t)x(t) + D(t)u(t), \tag{1.4c}$$

where $\mathcal{S} \subset [0, +\infty) \times \mathbb{R}^n$ is called the resetting set. In particular, (1.4a) describes the continuous-time dynamics of the IDLS, while (1.4b) represents the resetting law. For a particular trajectory $x(\cdot)$, we denote by t_k, $k \in \mathbb{N}$, the kth instant of time at which $(t, x(t))$ intersects $\mathcal{S}$; the instants t_k, $k = 1, \ldots, \mathbb{N}$, are called resetting times. According to the resetting law (1.4b), system (1.4a)–(1.4b) exhibits a finite jump from $x(t_k)$ to $x(t_k^+)$ at each resetting time t_k since, in general, $x(t_k^+) \neq x(t_k)$.

IDLSs can be either time-dependent (TD-IDLS), when the state jumps are time-driven, or state-dependent (SD-IDLS), when the state jumps occur when the trajectory reaches the resetting set $\mathcal{S}$. An example which falls into this category of systems is the automatic gear-box in cruise control (for more details and further examples, see [74, 76]).

Note that systems in the form (1.4a)–(1.4b) recover as particular cases CT-LTV systems when $A_d = 0$.

Chapter 9, where the robustness issues for IDLSs are discussed, ends Part II of the book.

For self-containedness purposes, the proofs of the main theorems are provided. Whenever possible, we give alternative (and simpler) proofs of those ones available in the literature, while maintaining a rigorous treatment of the matter; in any case, a reference is made to the paper where the theorem has been originally stated. Moreover, each chapter is equipped with a summary that recalls the main topics we have dealt with, outlines a brief history of the development of the research concerning such topics, and provides further references for alternative approaches other than the ones considered in the book.

Finally, all numerical computations done in the examples have been performed with the aid of the MATLAB software and the YALMIP parser [70].

Part I
Linear Systems

Chapter 2
FTS Analysis of Continuous-Time Linear Systems

2.1 Introduction

This chapter deals with the FTS analysis of CT-LTV systems. The first result of the chapter is a necessary and sufficient condition for FTS. Such a condition requires the computation of the state transition matrix of the system; for this reason, it is of little usefulness in practice, both for the intrinsic difficulty of computing the state transition matrix of a linear time-varying system and, more importantly, because it cannot be used in the design context.

Then, by using an approach based on time-varying quadratic Lyapunov functions, two further necessary and sufficient conditions for FTS are stated. The former requires the existence of a feasible solution to a certain DLMI; the latter involves a DLE.

It is shown that the DLE-based condition is more efficient from the computational point of view; however, DLMIs are useful in the context of the design problem. Indeed, in Chap. 3, by the latter approach we shall solve both the state and output feedback finite-time stabilization problems.

At the end of the chapter, the proposed approaches will be used to study the FTS properties of the car suspension system; such an engineering example will be continued in the following chapters to illustrate the effectiveness of the proposed FTS design techniques.

2.2 Problem Statement

In the following, we restrict Definition 1.1 of FTS to the case of CT-LTV systems, and we assume that both the initial set and the trajectory set are ellipsoids. Moreover, we assume that the trajectory set is a *time-varying* ellipsoid. All time-varying matrices, unless otherwise specified, are assumed to be bounded and piecewise continuous functions of time.

F. Amato et al., *Finite-Time Stability and Control*,
Lecture Notes in Control and Information Sciences 453,
DOI 10.1007/978-1-4471-5664-2_2, © Springer-Verlag London 2014

Definition 2.1 (FTS of CT-LTV Systems with Time-Varying Ellipsoidal Domains)
Given an initial time t_0, a positive scalar T, a positive definite matrix R, and a
positive definite matrix-valued function $\Gamma(\cdot)$, defined over $[t_0, t_0 + T]$, such that
$\Gamma(t_0) < R$, the time-varying linear system

$$\dot{x}(t) = A(t)x(t), \quad x(t_0) = x_0, \tag{2.1}$$

where $A(t) \in \mathbb{R}^{n \times n}$, is said to be finite-time stable with respect to $(t_0, T, R, \Gamma(\cdot))$ if

$$x_0^T R x_0 \le 1 \Rightarrow x(t)^T \Gamma(t)x(t) < 1, \quad t \in [t_0, t_0 + T]. \tag{2.2}$$

$\Diamond$

Remark 2.1 The assumption $\Gamma(t_0) < R$ guarantees that the closed ellipsoid

$$\left\{ x_0 : x_0^T R x_0 \le 1 \right\}$$

is a subset of the open ellipsoid $\{x_0 : x_0^T \Gamma(t_0)x_0 < 1\}$; this in turn guarantees the
well-posedness of Definition 2.1. $\Diamond$

2.3 Main Results

The following theorem gives a number of necessary and sufficient conditions for the
FTS of system (2.1).

Theorem 2.1 *The following statements are equivalent:*

(i) *System (2.1) is finite-time stable with respect to $(t_0, T, R, \Gamma(\cdot))$.*
(ii) *For all $t \in [t_0, t_0 + T]$,*

$$\Phi(t, t_0)^T \Gamma(t)\Phi(t, t_0) < R,$$

where $\Phi(t, t_0)$ is the state transition matrix of system (2.1).
(iii) *The matrix-valued function $W(\cdot) : [t_0, t_0 + T] \mapsto \mathbb{R}^{n \times n}$, solution of the DLE*

$$-\dot{W}(t) + A(t)W(t) + W(t)A^T(t) = 0, \quad t \in [t_0, t_0 + T], \tag{2.3a}$$

$$W(t_0) = R^{-1}, \tag{2.3b}$$

is positive definite and satisfies

$$C(t)W(t)C^T(t) < I, \quad t \in [t_0, t_0 + T], \tag{2.4}$$

where $C(\cdot)$ is any nonsingular matrix-valued function such that $\Gamma(t) = C^T(t)C(t)$ for $t \in [t_0, t_0 + T]$.

(iv) *Either one of the following inequalities holds*:

$$\lambda_{\max}\left[C(t)W(t)C^T(t)\right] < 1,   \qquad (2.5a)$$

$$\lambda_{\min}\left[C^{-T}(t)M(t)C^{-1}(t)\right] > 1,   \qquad (2.5b)$$

where $W(\cdot)$ is the positive definite solution of (2.3a)–(2.3b), *and $M(\cdot)$ is the positive definite solution of*

$$\dot{M}(t) + A^T(t)M(t) + M(t)A(t) = 0, \quad t \in [t_0, t_0 + T],   \qquad (2.6a)$$

$$M(t_0) = R,   \qquad (2.6b)$$

where $C(\cdot)$ is a nonsingular matrix-valued function such that $\Gamma(t) = C^T(t)C(t)$ for $t \in [t_0, t_0 + T]$.

(v) *The DLMI with terminal and initial conditions*

$$\dot{P}(t) + A^T(t)P(t) + P(t)A(t) < 0, \quad t \in [t_0, t_0 + T],   \qquad (2.7a)$$

$$P(t) > \Gamma(t), \quad t \in [t_0, t_0 + T],   \qquad (2.7b)$$

$$P(t_0) < R,   \qquad (2.7c)$$

admits a piecewise continuously differentiable symmetric solution $P(\cdot)$ defined over $[t_0, t_0 + T]$.   □

Before proving the theorem, some comments are in order.

Remark 2.2 The equivalence of (i) and (ii) has been proved in [10]. The fact that condition (v) implies FTS of system (2.1) dates back to the same work; however, the equivalence of the two conditions and condition (iii) was only proven in the paper [27]. Finally, the proof of the equivalence of (iv) and FTS can be found in [49]. The proof of the theorem presented here follows a slight different machinery, in order to compact all results together.   ◊

Remark 2.3 Condition (ii) is hard to apply in the time-varying case, as it requires the computation of the state transition matrix. There is also another important drawback with Condition (ii); indeed, it is simple to recognize that it is not useful for design purposes.   ◊

Remark 2.4 In Sect. 2.4, it will be shown that conditions (iii) and (iv), which require the solution of a DLE, are more efficient from the computational point of view in comparison with condition (v) (a similar discussion can be found in [26]). However, (2.7a)–(2.7c) will be the starting point for the solution of the design problem; indeed, the inequality allows us to put the controller design in the convex optimization framework.   ◊

Proof $\boxed{(i) \Leftrightarrow (ii)}$ Let x_0 be such that $x_0^T R x_0 \leq 1$. Then

$$x^T(t)\Gamma(t)x(t) = x_0^T \Phi^T(t,t_0)\Gamma(t)\Phi(t,t_0)x_0$$

$$< x_0^T R x_0 \leq 1. \tag{2.8}$$

Therefore, system (2.25) is finite-time stable.

The necessity is shown by contradiction. Let us assume that for some $\bar{t}$ and $\bar{x}$,

$$\bar{x}^T \Phi^T(t,t_0)\Gamma(t)\Phi(t,t_0)\bar{x} \geq \bar{x}^T R\bar{x}. \tag{2.9}$$

Now let $x_0 = \lambda\bar{x}$, where λ is such that $x_0^T R x_0 = 1$. Then (2.9) implies that

$$x_0^T \Phi^T(t,t_0)\Gamma(t)\Phi(t,t_0)x_0 \geq 1. \tag{2.10}$$

Therefore,

$$x(\bar{t})^T \Gamma(t)x(\bar{t}) = x_0^T \Phi^T(t,t_0)\Gamma(t)\Phi(t,t_0)x_0 \geq 1, \tag{2.11}$$

which contradicts the initial assumption that system (2.25) is finite-time stable.

In the sequel, we will first prove that (i) $\Leftrightarrow$ (iii) and (iii) $\Leftrightarrow$ (iv); then we show that (iii) $\Rightarrow$ (v) and (v) $\Rightarrow$ (i), which completes the proof.

$\boxed{(i) \Leftrightarrow (iii)}$ First, note that, given a real symmetric and positive definite matrix-valued function $\Gamma(\cdot) \in \mathbb{R}^n$, it is always possible to find a nonsingular matrix-valued function $C(\cdot) : [t_0, t_0 + T] \mapsto \mathbb{R}^{n \times n}$ such that

$$\Gamma(t) = C^T(t)C(t), \quad t \in [t_0, t_0 + T].$$

Since (i) $\Leftrightarrow$ (ii), it follows that system (2.1) is finite-time stable if and only if

$$\Phi^T(t,t_0)\Gamma(t)\Phi(t,t_0) - R < 0, \quad t \in [t_0, t_0 + T],$$

which is equivalent to

$$\Phi^T(t,t_0)C^T(t)C(t)\Phi(t,t_0) - R < 0$$

$$\Leftrightarrow R^{-\frac{1}{2}}\Phi^T(t,t_0)C^T(t)C(t)\Phi(t,t_0)R^{-\frac{1}{2}} - I < 0$$

$$\Leftrightarrow C(t)\Phi(t,t_0)R^{-1}\Phi^T(t,t_0)C^T(t) - I < 0 \tag{2.12}$$

for all $t \in [t_0, t_0 + T]$. Now, if we let

$$W(t) = \Phi(t,t_0)R^{-1}\Phi^T(t,t_0),$$

then (2.12) can be rewritten as (2.4).

Furthermore, taking into account that the transition matrix $\Phi(t, t_0)$ solves the matrix-valued differential equation

$$\frac{\partial}{\partial t}\Phi(t, t_0) = A(t)\Phi(t, t_0), \quad t \in [t_0, t_0 + T], \quad \Phi(t_0, t_0) = I,$$

it readily follows that $W(\cdot)$ is a positive definite solution of (2.3a)–(2.3b).

$\boxed{\text{(iii)} \Leftrightarrow \text{(iv)}}$ First, note that from the definition of $W(\cdot)$ and $C(\cdot)$ it readily follows that inequality (2.4) is equivalent to (2.5a).

Furthermore, the equivalence of (2.5a) and (2.5b) follows by letting

$$M(\cdot) = W^{-1}(\cdot)$$

over $[t_0, t_0 + T]$.

$\boxed{\text{(iii)} \Rightarrow \text{(v)}}$ We have shown that if $W(\cdot)$ satisfies (2.3a)–(2.3b) and (2.4), then system (2.1) is finite-time stable. By continuity arguments, if system (2.1) is finite-time stable wrt $(t_0, T, R, \Gamma(\cdot))$, then there exists a real scalar ε such that also the following system

$$\dot{z}(t) = \left(A(t) + \frac{\varepsilon}{2}I\right)z(t), \quad z(t_0) = x_0,$$

is finite-time stable with respect to $(t_0, T, R, \Gamma(\cdot))$.

Taking again into account the equivalence of conditions (i) and (iii), we denote by $W_\varepsilon(\cdot)$ the continuous positive definite matrix-valued solution of

$$-\dot{W}_\varepsilon(t) + A(t)W_\varepsilon(t) + W_\varepsilon(t)A^T(t) + \varepsilon W_\varepsilon(t) = 0, \quad t \in [t_0, t_0 + T], \quad (2.13\text{a})$$

$$W_\varepsilon(t_0) = R^{-1}, \quad (2.13\text{b})$$

which also satisfies

$$C(t)W_\varepsilon(t)C^T(t) < I, \quad t \in [t_0, t_0 + T].$$

Exploiting continuity arguments once more, it turns out that there exists a real scalar $\alpha > 1$ such that

$$\alpha C(t)W_\varepsilon(t)C^T(t) < I, \quad t \in [t_0, t_0 + T]. \quad (2.14)$$

Let $X(t) = \alpha W_\varepsilon(t)$, $t \in [t_0, t_0 + T]$; inequality (2.14) can be rewritten as

$$C(t)X(t)C^T(t) < I, \quad t \in [t_0, t_0 + T]. \quad (2.15)$$

Since $\dot{X}(t) = \alpha \dot{W}_\varepsilon(t)$, from (2.13a) we obtain

$$-\dot{X}(t) + A(t)X(t) + X(t)A^T(t) + \varepsilon X(t) = 0, \quad t \in [t_0, t_0 + T].$$

Taking into account the positive definitiveness of $X(t)$, it follows that

$$-\dot{X}(t) + A(t)X(t) + X(t)A^T(t) < 0 \qquad (2.16)$$

for $t \in [t_0, t_0 + T]$. Furthermore, taking into account (2.13b), we obtain

$$X(t_0) > R^{-1}. \qquad (2.17)$$

Eventually, letting

$$P(t) = X^{-1}(t), \quad t \in [t_0, t_0 + T],$$

inequalities (2.7a)–(2.7c) can be easily obtained from (2.15)–(2.17).

$\boxed{(v) \Rightarrow (i)}$ Let us consider $V(t,x) = x^T(t)P(t)x(t)$. Given a system trajectory $x(\cdot)$, the time derivative of $V(t,x)$ reads

$$\dot{V}(t,x) = x^T(t)\big(\dot{P}(t) + A(t)^T P(t) + P(t)A(t)\big)x(t),$$

which is negative definite by virtue of (2.7a). This implies that $V(t,x)$ is strictly decreasing along the trajectories of system (2.1). Now, if we consider an initial state x_0 such that $x_0^T R x_0 \le 1$, we have, for all $t \in [t_0, t_0 + T]$,

$$x(t)^T \Gamma(t)x(t) < x(t)^T P(t)x(t) \quad \text{by (2.7b)}$$

$$< x_0^T P(t_0)x_0$$

$$< x_0^T R x_0 \le 1 \quad \text{by (2.7c)},$$

which implies that system (2.1) is finite-time stable with respect to $(t_0, T, R, \Gamma(\cdot))$. $\diamond$

2.4 Computational Issues

The numerical example considered in this section is introduced to discuss the effectiveness and some computational issues related to the necessary and sufficient conditions for the FTS of CT-LTV systems proposed in Theorem 2.1.

Let us first consider the second-order LTV system

$$\dot{x}(t) = \begin{pmatrix} 0.4 \cdot t & 1 \\ -1 & -1 + 0.4 \cdot t \end{pmatrix} x(t) \qquad (2.18)$$

and the time interval $[0, 1]$. We set

$$R = \begin{pmatrix} 2.5 & 0 \\ 0 & 2.5 \end{pmatrix}, \quad \Gamma = \begin{pmatrix} \gamma & 0 \\ 0 & \gamma \end{pmatrix}, \qquad (2.19)$$

Table 2.1 Values of γ_{max} satisfying Theorem 2.1 for the LTV system (2.18)

Condition	Sample time (T_s) [s]	γ_{max}	Average computation time for a single iteration [s]
DLMI (2.7a)–(2.7c)	0.1	1.9226	2.0
	0.05	1.9321	12.3
	0.025	1.9367	139.6
Solution of (2.3a)–(2.3b) and check of inequality (2.4)	$2 \cdot 10^{-4}$	1.9402	1.1

and we exploit the necessary and sufficient conditions (iii) and (v) of Theorem 2.1 in a linear search, in order to estimate the maximum value of the parameter γ, say γ_{max}, such that system (2.18) is finite-time stable wrt $(0, 1, R, \Gamma)$.

Note that, in order to recast the DLMI condition (2.7a)–(2.7c) in terms of LMIs, the matrix-valued function $P(\cdot)$ has been assumed piecewise linear by dividing the time interval $[0, T]$ in $n = T/T_s$ subintervals and assuming the time derivative of $P(\cdot)$ constant in each subinterval:

$$P(t) = \begin{cases} P_0 + \Theta_1(t - t_0), & t \in [t_0, t_0 + T_s], \\ P_0 + \sum_{h=1}^{j} \Theta_h T_s + \Theta_{j+1}(t - jT_s - t_0), & t \in (t_0 + jT_s, t_0 + (j+1)T_s], \\ j = 1, \ldots, J, \end{cases}$$

where $J = \max\{j \in \mathbb{N} : j < T/T_s\}$, $T_s \ll T$, and P_0, Θ_l, $l = 1, \ldots, J + 1$, are the optimization variables. It is straightforward to recognize that such a piecewise function can approximate a generic continuous $P(\cdot)$ with adequate accuracy, provided that the length of T_s is sufficiently small.

The obtained estimates of γ_{max}, the corresponding values T_s, and the average computation times for a single iteration of the linear search are shown in Table 2.1. These results have been obtained by running the Matlab LMI Toolbox [48] on a PC equipped with an Intel i5-460M processor and 6 GB of RAM.

The example shows that the condition based on the solution of the DLE is more efficient than the DLMI-based approach, both in terms of computing time and in terms of estimation of γ_{max}.

It is clear from Table 2.1 that, when the DLMIs are involved, increasing the sampling time causes only a marginal improvement in terms of the γ_{max} estimation, while the time efficiency decreases dramatically.

2.5 Application to the Car Suspension System

In this section, in order to illustrate the proposed technique, we present a typical engineering case study, namely a vehicle active suspension system.

Fig. 2.1 Schematic
representation of the active
suspension system

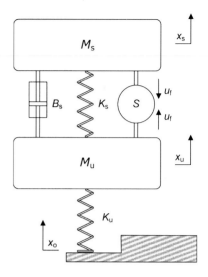

2.5.1 The Model

The model considered to this aim is taken from [40], where the authors focus on the design of an $\mathcal{H}_\infty$ control scheme satisfying some output and control constraints.

    The scheme of a two-degrees-of-freedom (2-DOF) quarter-car model is reported in Fig. 2.1: the system comprises the sprung mass M_s, the unsprung mass M_u, the suspension damper with damping coefficient B_s, the suspension spring with elastic coefficient K_s, the elastic effect caused by the tire deflection, modeled by means of a spring with elastic coefficient K_u, and the hydraulic actuator S generating a scalar active force u_f. As state variables, we choose the suspension stroke $x_s - x_u$, the tire deflection $x_u - x_o$, and the derivatives with respect to time of x_s and x_u, that is,

$$x_1 = x_s - x_u,$$

$$x_2 = \dot{x}_s,$$

$$x_3 = x_u - x_o,$$

$$x_4 = \dot{x}_u,$$

where x_s and x_u are the vertical displacements of the sprung and unsprung masses, respectively, and x_o is the vertical ground displacement caused by the road unevenness. The resulting open-loop dynamical model reads

$$\dot{x}(t) = \begin{bmatrix} 0 & 1 & 0 & -1 \\ -\dfrac{K_s}{M_s} & -\dfrac{B_s}{M_s} & 0 & \dfrac{B_s}{M_s} \\ 0 & 0 & 0 & 1 \\ \dfrac{K_s}{M_u} & \dfrac{B_s}{M_u} & -\dfrac{K_u}{M_u} & -\dfrac{B_s}{M_u} \end{bmatrix} x(t) + \begin{bmatrix} 0 \\ \dfrac{u_{\max}}{M_s} \\ 0 \\ -\dfrac{u_{\max}}{M_u} \end{bmatrix} u(t) + \begin{bmatrix} 0 \\ 0 \\ -1 \\ 0 \end{bmatrix} w(t),$$

$$(2.20)$$

where the normalized active force $u = u_f/u_{max}$ is the control input, and $w = \dot{x}_o$ is an exogenous disturbance generated by the road roughness. The values of the model parameters used in this example are

$$M_s = 320 \text{ kg}, \quad K_s = 50 \frac{\text{kN}}{\text{m}}, \quad B_s = 2000 \frac{\text{N s}}{\text{m}},$$

$$K_u = 200 \frac{\text{kN}}{\text{m}}, \quad M_u = 40 \text{ kg}, \quad u_{max} = 100 \text{ kN}.$$

2.5.2 FTS Analysis

For the system illustrated above, it is important to prevent excessive suspension bottoming, which can lead to a rapid deterioration of ride comfort and possible structural damage. This goal can be translated into a constraint on the suspension stroke limitation in the form

$$|x_1| = |x_s(t) - x_u(t)| \le SS_{max}, \quad t \ge 0, \tag{2.21}$$

where SS_{max} is the maximum allowable suspension stroke.

For simplicity, here we will limit our scope to the case of road bumps (or holes): in this case, the vertical ground displacement can be approximated to a step function, and the disturbance input $w(t)$ is impulsive. Therefore, the bump effect can be reduced to an instantaneous change of the initial value of x_3, and the disturbance term is identically zero for all $t > 0$ (we set $t = 0$ as the instant when the tyre hits the bump).

This problem can be readily cast in the framework of FTS: first, let us characterize the family of bumps for which we want to perform the FTS analysis by defining an ellipsoidal constraint on the admissible initial conditions; assuming that the system is at steady state at $t = 0^-$, that is, all the state variables but x_3 are zero, the weighting matrix would ideally read

$$R = \text{diag}\left(\infty \quad \infty \quad \frac{1}{x_{3max}^2} \quad \infty\right). \tag{2.22}$$

This choice of R considers all the possible disturbances yielding an instantaneous change of $|x_3| \le x_{3\,max}$.

The suspension stroke constraint can be taken into account through the choice of a suitable weighting matrix function $\Gamma(\cdot)$. The matrix

$$\Gamma(t) = \text{diag}\left(\frac{1}{(x_{1\,max}e^{-\frac{t}{\tau}})^2} \quad 0 \quad 0 \quad 0\right) \tag{2.23}$$

defines an exponentially decaying threshold on the suspension stroke; for the initial maximum value, we set $x_{1\,max} = 0.02$ m and $\tau = 0.4$ s.

Fig. 2.2 Time history of the squared weighted norm of the state of system (2.20); free evolution from initial condition (0 0 0.08 0)

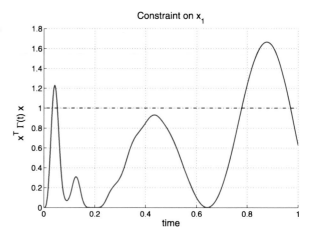

With the given choices of R and $\Gamma(\cdot)$, we can now tackle the problem of analyzing the suspension stroke constraint by solving the FTS analysis problem for system (2.20) (with $w \equiv 0$) with respect to $(0, 1, R, \Gamma(\cdot))$.

In order to avoid numerical singularity issues, the above ideal values of the weighting matrices cannot be used. Instead, we relax the constraints by setting the ∞ entries of R to very large values and the zero entries of $\Gamma(\cdot)$ to very small values.

Figure 2.2 reports the time evolution of $x^T(t)\Gamma(t)x(t)$ for $x_3(0) = 0.08$ m, showing the violation of the constraint on x_1. The corresponding state response is reported in Fig. 2.3.

Applying condition (iii) of Theorem 2.1, it can be easily checked that the system under consideration is not finite-time stable wrt $(0, 1, R, \Gamma(\cdot))$ when $x_{3\,\text{max}} = 0.08$, which is consistent with the simulation results shown above.

Now, we can exploit the analysis conditions to find the maximum value of $x_3(0)$ that does not yield a violation of the assigned suspension stroke constraint. To this aim, it is sufficient to iteratively apply Theorem 2.1 changing the values of $x_{3\,\text{max}}$ in (2.22). By means of a dichotomous search it is readily found that such a value is $\bar{x}_3 = 0.049$ m. This result is confirmed by the state response depicted in Fig. 2.4.

2.6 Summary

In this chapter we have discussed the FTS problem for the class of CT-LTV systems. To this end, a methodology based on the use of time-varying quadratic Lyapunov functions has been developed.

The main result of the section consists of some necessary and sufficient conditions for FTS. The first condition, which involves the state transition matrix, is hard from the computational point of view (unless simple cases are considered); also, it is not useful for design purposes. Therefore, condition (ii) in Theorem 2.1 is not practical.

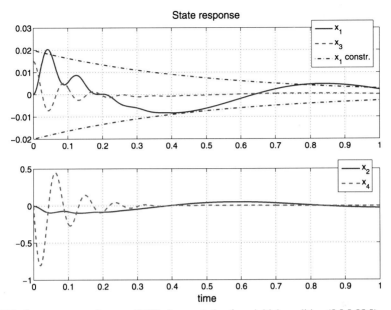

Fig. 2.3   State response of system (2.20): free evolution from initial condition (0 0 0.08 0)

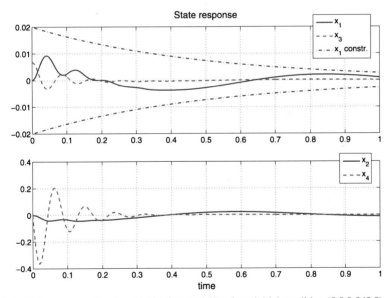

Fig. 2.4   State response of system (2.20): free evolution from initial condition (0 0 0.049 0)

Condition (v) in Theorem 2.1 allows us to check FTS of system 2.1 via a DLMI coupled with a pair of (time-varying) LMIs. In the former papers [10, 19], the sufficiency of the condition was only proved. In a later work (see [27]), the necessity was demonstrated, together with the equivalence with condition (iii) in Theorem 2.1, which, differently from (v), requires the solution of a DLE. The latter condition is more efficient from the computational point of view, as it is shown in the example section. Finally, it is equivalent to Condition (iv), which was independently derived by Garcia et al. [49].

As it will be shown in the following chapter, the DLMI-based condition plays an important role in the derivation of the design conditions. Indeed, the optimization of the controller matrices requires the availability of a convex condition to be numerically implemented.

The methodology has been applied to design a dynamical controller for the car suspension system derived in Sect. 3.4. This engineering example gave also the opportunity of illustrating some approaches for the numerical solution of DLEs and DLMIs and for showing that DLEs are more efficient from the computational point of view than DLMIs.

By restricting Definition 1.1 to the case of continuous-time linear time-invariant (CT-LTI) systems and assuming that both the initial set and the trajectory set are time-invariant ellipsoids having the same shape, a different approach for the FTS study is proposed in [6]. To this end, consider a system of the form

$$\dot{x}(t) = Ax(t), \tag{2.24}$$

where $A \in \mathbb{R}^{n \times n}$. For CT-LTI systems, the FTS definition (2.1) particularizes as follows.

Definition 2.2 (FTS of CT-LTI Systems with Ellipsoidal Domains) Given a positive scalar T and positive definite matrices R and Γ with $\Gamma > R$, the CT-LTI system

$$\dot{x}(t) = Ax(t), \quad x(0) = x_0, \tag{2.25}$$

is said to be finite-time stable with respect to (T, R, Γ) if

$$x_0^T R x_0 \leq 1 \Rightarrow x(t)^T \Gamma x(t) < 1, \quad t \in [0, T]. \tag{2.26}$$

$\diamond$

The following theorem is derived under the assumption that the initial set and the trajectory set have the same shape, namely $\Gamma = \rho R$, for a given positive scalar $\rho < 1$.

Theorem 2.2 [6] *System (2.25) is finite-time stable with respect to $(T, R, \rho R)$ with $0 < \rho < 1$ if letting $\tilde{Q} = R^{-\frac{1}{2}} Q R^{-\frac{1}{2}}$, there exist a nonnegative scalar α and a positive definite matrix $Q \in \mathbb{R}^{n \times n}$ such that*

$$A\tilde{Q} + \tilde{Q}A^T - \alpha \tilde{Q} < 0, \tag{2.27a}$$

$$\text{cond}(Q) < \frac{1}{\rho} e^{-\alpha T}, \tag{2.27b}$$

where $\text{cond}(Q) = \frac{\lambda_{\max}(Q)}{\lambda_{\min}(Q)}$ denotes the condition number of Q.   □

To prove Theorem 2.2, consider the Lyapunov function $V(x) := x^T \tilde{Q}^{-1} x$. Suppose that

$$\dot{V}\big(x(t)\big) < \alpha V\big(x(t)\big) \tag{2.28}$$

for all $t \in [0, T]$. The first step consists of proving that conditions (2.28) and (2.27b) imply that system (2.25) is finite-time stable with respect to $(T, R, \rho R)$. Indeed, by dividing both sides of (2.28) by $V(x)$, and integrating from 0 to t, with $t \in (0, T]$, we obtain

$$\log \frac{V(x(t))}{V(x(0))} < \alpha t. \tag{2.29}$$

We have, for $t \in (0, T]$,

$$x^T(t) R^{1/2} Q^{-1} R^{1/2} x(t) \geq \lambda_{\min}\big(Q^{-1}\big) x^T(t) R x(t) \tag{2.30}$$

and

$$x^T(0) R^{1/2} Q^{-1} R^{1/2} x(0) e^{\alpha t} \leq \lambda_{\max}\big(Q^{-1}\big) e^{\alpha T}. \tag{2.31}$$

Putting together (2.29)–(2.31), we have

$$x^T(t) R x(t) < \frac{\lambda_{\max}(Q)}{\lambda_{\min}(Q)} e^{\alpha T} = \frac{1}{\rho}. \tag{2.32}$$

To conclude the proof, note that condition (2.28) is equivalent to (2.27a).

In Theorem 2.2 an exponential time weighting $e^{-\alpha t/2}$ of the state with a *negative* exponent is used, while in the Lyapunov stability context one uses an exponential time weighting with a positive exponent. This allows us to relax the classical conditions for asymptotic stability and to establish the FTS of the system under consideration even if it is not asymptotically stable.

It is easy to check that condition (2.27b) can be turned into the LMI in the matrix variable Q

$$\rho e^{\alpha T} I < Q < I.$$

Therefore, from a computational point of view, it is important to notice that, once a value for α has been fixed, the feasibility of the conditions stated in Theorem 2.2 can be turned into an LMI-based feasibility problem.

As said, one strong drawback of the approach followed in Theorem 2.2 is that we have to restrict the matrix Γ to be in the form $\Gamma = \rho R$ with $\rho < 1$; in other words, the (time-invariant) trajectory set is constrained to be a scaled version of the initial set. This is a weakness with respect to the approach followed in Theorem 2.1. However, Theorem 2.2 plays an important role in the following developments since it probably represents the first attempt to applying an LMI-based approach to solve FTS problems.

Conditions for finite-time stabilization via state and output feedback, consistent with Definition 2.2, are proposed in [6] and [12], respectively, and will be briefly illustrated in Sect. 3.5.

Chapter 3
Controller Design for the Finite-Time Stabilization of Continuous-Time Linear Systems

3.1 Introduction

In this chapter, we consider the finite-time stabilization problem for CT-LTV systems. The necessary and sufficient condition stated in Theorem 2.1 is the starting point for the derivation of a necessary and sufficient condition for finite-time stabilization via state feedback; such condition requires the existence of a feasible solution to a certain DLMI.

Then we discuss the finite-time stabilization problem via dynamical output feedback. To this end, we use the nonlinear change of matrix variables proposed in [47], in order to obtain explicit necessary and sufficient conditions for the computation of controller matrices. Such results require the feasibility of a certain DLMI subject to an initial condition, which can be converted to an LMI only at the price of some conservatism.

In the example section at the end of the chapter, the proposed techniques are applied to the design of a controller for the active suspension system introduced in Sect. 2.5.

3.2 Problem Statement

In the following, the main stabilization problems we deal with in this chapter are precisely stated. All the involved time-varying matrices, unless otherwise stated, are assumed to be bounded, piecewise continuous functions of time.

Problem 3.1 (Finite-Time Stabilization of CT-LTV Systems Via State Feedback) Let us consider the time-varying linear system

$$\dot{x}(t) = A(t)x(t) + B(t)u(t), \quad x(t_0) = x_0, \tag{3.1}$$

where $A(t) \in \mathbb{R}^{n \times n}$, $B(t) \in \mathbb{R}^{n \times m}$, and $u(t) \in \mathbb{R}^m$ is the control input. Then, given an initial time t_0, a positive scalar T, a positive definite matrix R, and a positive

F. Amato et al., *Finite-Time Stability and Control*,
Lecture Notes in Control and Information Sciences 453,
DOI 10.1007/978-1-4471-5664-2_3, © Springer-Verlag London 2014

definite matrix-valued function $\Gamma(\cdot)$, defined over $[t_0, t_0 + T]$, such that $\Gamma(t_0) < R$, find a state feedback controller of the form

$$u(t) = K(t)x(t) \tag{3.2}$$

such that the closed-loop system obtained by the connection of (3.1) and (3.2), namely

$$\dot{x}(t) = \big(A(t) + B(t)K(t)\big)x(t), \quad x(t_0) = x_0, \tag{3.3}$$

is finite-time stable with respect to $(t_0, T, R, \Gamma(\cdot))$. $\Diamond$

The next problem deals with the output feedback finite-time stabilization; note that dynamical controllers with the same order of the plant are considered and that we assume, for the sake of simplicity, that the weighting matrix of the state of the closed-loop system does not contain cross-coupling terms between the system state and the controller state (on the other hand, there is no physical meaning in introducing such cross-coupling terms).

Problem 3.2 (Finite-Time Stabilization of CT-LTV Systems Via Output Feedback) Consider the time-varying linear system

$$\dot{x}(t) = A(t)x(t) + B(t)u(t), \quad x(t_0) = x_0, \tag{3.4a}$$

$$y(t) = C(t)x(t), \tag{3.4b}$$

where $C(t) \in \mathbb{R}^{r \times n}$, and $y(t)$ is the output. Then, given an initial time t_0, a positive scalar T, two positive definite matrices R, R_K and two positive definite symmetric matrix-valued functions $\Gamma(\cdot)$, $\Gamma_K(\cdot)$, defined over $[t_0, t_0 + T]$, such that $\Gamma(t_0) < R$, $\Gamma_K(t_0) < R_K$, find a dynamical output feedback controller of the form

$$\dot{x}_c(t) = A_K(t)x_c(t) + B_K(t)y(t), \quad x_c(t_0) = 0, \tag{3.5a}$$

$$u(t) = C_K(t)x_c(t) + D_K(t)y(t), \tag{3.5b}$$

where $A_K(t) \in \mathbb{R}^{n \times n}$, $B_K(t) \in \mathbb{R}^{n \times r}$, $C_K(t) \in \mathbb{R}^{m \times n}$, and $D_K(t) \in \mathbb{R}^{m \times r}$, such that the closed-loop system obtained by the connection of (3.4a)–(3.4b) and (3.5a)–(3.5b) is finite-time stable with respect to

$$\big(t_0, T, \operatorname{diag}(R, R_K), \operatorname{diag}\big(\Gamma(\cdot), \Gamma_K(\cdot)\big)\big).   \Diamond$$

3.3 Main Results

Starting from condition (v) in Theorem 2.1, the following necessary and sufficient condition to solve Problem 3.1 is provided. A preliminary version of this result, accounting only for the sufficiency of the condition, has appeared in [11, 19].

Theorem 3.1 *Problem* 3.1 *is solvable if and only if there exist a piecewise continuously differentiable symmetric matrix-valued function* $Q(\cdot)$ *and a continuous matrix-valued function* $L(\cdot)$ *that satisfy the following DLMI problem with initial and terminal conditions*:

$$- \dot{Q}(t) + A(t)Q(t) + Q(t)A^T(t) + L^T(t)B^T(t) + B(t)L(t) < 0,$$

$$t \in [t_0, t_0 + T], \tag{3.6a}$$

$$Q(t) < \Gamma^{-1}(t), \quad t \in [t_0, t_0 + T], \tag{3.6b}$$

$$Q(t_0) > R^{-1}. \tag{3.6c}$$

The controller gain that solves Problem 3.1 *is*

$$K(t) = L(t)Q^{-1}(t). \qquad \square$$

Proof When the state feedback (3.2) is considered, the necessary and sufficient condition (v) of Theorem 2.1 applied to the closed-loop system requires the existence of a piecewise continuously differentiable matrix $P(\cdot)$ such that

$$\dot{P}(t) + \big(A(t) + B(t)K(t)\big)^T P(t) + P(t)\big(A(t) + B(t)K(t)\big) < 0,$$

$$t \in [t_0, t_0 + T], \tag{3.7a}$$

$$P(t) > \Gamma(t), \quad t \in [t_0, t_0 + T], \tag{3.7b}$$

$$P(t_0) < R. \tag{3.7c}$$

Let $Q(t) = P^{-1}(t)$ and pre- and post-multiply (3.7a) by $Q(t)$. Taking into account that $\dot{Q}(t) = -P^{-1}(t)\dot{P}(t)P^{-1}(t)$, we obtain

$$- \dot{Q}(t) + Q(t)\big(A(t) + B(t)K(t)\big)^T + \big(A(t) + B(t)K(t)\big)Q(t) < 0,$$

$$t \in [t_0, t_0 + T], \tag{3.8a}$$

$$Q(t) < \Gamma^{-1}(t), \quad t \in [t_0, t_0 + T], \tag{3.8b}$$

$$Q(t_0) > R^{-1}. \tag{3.8c}$$

The proof of the theorem readily follows by letting, according to [51], $L(t) = K(t)Q(t)$. $\Diamond$

Note that, even if system (3.4a)–(3.4b) is time-invariant, the feedback gain turns out to be time-varying. This is typical for finite-horizon control techniques, such as the linear quadratic optimal control [30] and H_∞ control [32].

Before introducing a necessary and sufficient condition to solve Problem 3.2, we recall the following lemma, which will be useful in the proof of the next result.

Lemma 3.1 [33], [1, Chap. 5] *Given symmetric matrices $S \in \mathbb{R}^{n \times n}$ and $Q \in \mathbb{R}^{n \times n}$, the following statements are equivalent:*

(i) *There exist symmetric matrices $U \in \mathbb{R}^{n \times n}$ and $T \in \mathbb{R}^{n \times n}$ and nonsingular matrices $M \in \mathbb{R}^{n \times n}$ and $N \in \mathbb{R}^{n \times n}$ such that*

$$P := \begin{pmatrix} S & M \\ M^T & U \end{pmatrix}, \qquad P^{-1} := \begin{pmatrix} Q & N \\ N^T & T \end{pmatrix}. \tag{3.9}$$

(ii)

$$\begin{pmatrix} Q & I \\ I & S \end{pmatrix} > 0. \tag{3.10}$$

$\square$

It is now possible to state the following necessary and sufficient condition for finite-time stabilization via output feedback, which represents the main result of this chapter. Again, a preliminary version of the theorem has appeared in the papers [11, 19], where only the sufficiency part is proven.

Theorem 3.2 *Problem 3.2 is solvable if and only if there exist piecewise continuously differentiable symmetric matrix-valued functions $Q(\cdot)$ and $S(\cdot)$, a continuous nonsingular matrix-valued function $N(\cdot)$, and continuous matrix-valued functions $\hat{A}_K(\cdot)$, $\hat{B}_K(\cdot)$, $\hat{C}_K(\cdot)$, and $D_K(\cdot)$ such that (the time argument is omitted for brevity)*

$$\begin{pmatrix} \Theta_{11} & \Theta_{12} \\ \Theta_{12}^T & \Theta_{22} \end{pmatrix} < 0, \quad t \in [t_0, t_0 + T], \tag{3.11a}$$

$$\begin{pmatrix} Q & \Psi_{12} & \Psi_{13} & \Psi_{14} \\ \Psi_{12}^T & \Psi_{22} & 0 & 0 \\ \Psi_{13}^T & 0 & I & 0 \\ \Psi_{14}^T & 0 & 0 & I \end{pmatrix} > 0, \quad t \in [t_0, t_0 + T], \tag{3.11b}$$

$$\begin{pmatrix} Q(t_0) & I \\ I & S(t_0) \end{pmatrix} < \begin{pmatrix} \Delta_{11} & Q(t_0)R \\ RQ(t_0) & R \end{pmatrix}, \tag{3.11c}$$

where

$$\Theta_{11} = -\dot{Q} + AQ + QA^T + B\hat{C}_K + \hat{C}_K^T B^T,$$

$$\Theta_{12} = A + \hat{A}_K^T + BD_K C,$$

$$\Theta_{22} = \dot{S} + SA + A^T S + \hat{B}_K C + C^T \hat{B}_K^T,$$

$$\Psi_{12} = I - Q\Gamma,$$

$$\Psi_{13} = Q\Gamma^{1/2},$$

$$\Psi_{14} = N\Gamma_K^{1/2},$$

$$\Psi_{22} = S - \Gamma,$$

$$\Delta_{11} = Q(t_0)RQ(t_0) + N(t_0)R_K N^T(t_0). \qquad \square$$

Proof The connection between systems (3.4a)–(3.4b) and (3.5a)–(3.5b) reads

$$\begin{pmatrix} \dot{x}(t) \\ \dot{x}_c(t) \end{pmatrix} = \begin{pmatrix} A(t) + B(t)D_K(t)C(t) & B(t)C_K(t) \\ B_K(t)C(t) & A_K(t) \end{pmatrix} \begin{pmatrix} x(t) \\ x_c(t) \end{pmatrix}$$

$$=: A_{\mathrm{CL}}(t)x_{\mathrm{CL}}(t). \tag{3.12}$$

Define $R_{\mathrm{CL}} = \mathrm{diag}(R, R_K)$ and $\Gamma_{\mathrm{CL}}(t) = \mathrm{diag}(\Gamma(t), \Gamma_K(t))$; according to condition v) of Theorem 2.1, we have that system (3.12) is finite-time stable wrt $(t_0, T, R_{\mathrm{CL}}, \Gamma_{\mathrm{CL}}(\cdot))$ if and only if there exists a piecewise continuously differentiable symmetric matrix-valued function $P(\cdot)$ such that

$$\dot{P}(t) + A_{\mathrm{CL}}^T(t)P(t) + P(t)A_{\mathrm{CL}}(t) < 0, \quad t \in [t_0, t_0 + T], \tag{3.13a}$$

$$P(t) > \Gamma_{\mathrm{CL}}(t), \quad t \in [t_0, t_0 + T], \tag{3.13b}$$

$$P(t_0) < R_{\mathrm{CL}}. \tag{3.13c}$$

Now let us define, according to Lemma 3.1,

$$P(t) = \begin{pmatrix} S(t) & M(t) \\ M^T(t) & U(t) \end{pmatrix}, \qquad P^{-1}(t) = \begin{pmatrix} Q(t) & N(t) \\ N^T(t) & \star \end{pmatrix}, \tag{3.14a}$$

$$\Pi_1(t) = \begin{pmatrix} Q(t) & I \\ N^T(t) & 0 \end{pmatrix}, \qquad \Pi_2(t) = \begin{pmatrix} I & S(t) \\ 0 & M^T(t) \end{pmatrix}. \tag{3.14b}$$

Note that, by definition,

$$S(t)Q(t) + M(t)N^T(t) = I, \tag{3.15a}$$

$$Q(t)\dot{S}(t)Q(t) + N(t)\dot{M}^T(t)Q(t) + Q(t)\dot{M}(t)N^T(t) + N(t)\dot{U}(t)N^T(t)$$

$$= -\dot{Q}(t), \tag{3.15b}$$

$$P(t)\Pi_1(t) = \Pi_2(t). \tag{3.15c}$$

By pre- and post-multiplying (3.13a)–(3.13c) by $\Pi_1^T(t)$ and $\Pi_1(t)$, respectively, taking into account (3.15a)–(3.15c) and Lemma 3.1, the proof follows once we let

(time is omitted for brevity)

$$\begin{pmatrix} Q(t) & I \\ I & S(t) \end{pmatrix} > 0, \tag{3.16a}$$

$$\hat{B}_K = MB_K + SBD_K, \tag{3.16b}$$

$$\hat{C}_K = C_K N^T + D_K CQ, \tag{3.16c}$$

$$\hat{A}_K = \dot{S}Q + \dot{M}N^T + MA_K N^T + SBC_K N^T + MB_K CQ$$
$$+ S(A + BD_K C)Q. \tag{3.16d}$$

Note that (3.16a) does not need to be explicitly imposed since it is implied by (3.11b). ◊

Remark 3.1 The conditions in Theorem 3.2 can be simplified. Indeed, note that the feasibility of (3.11a) is equivalent to the existence of $Q(\cdot)$, $S(\cdot)$, $\hat{B}_K(\cdot)$, and $\hat{C}_K(\cdot)$ such that

$$\Theta_{11}(t) < 0, \tag{3.17a}$$

$$\Theta_{22}(t) < 0. \tag{3.17b}$$

To understand this point, note that an admissible solution of (3.11a) also satisfies (3.17a)–(3.17b). Conversely, let us assume that (3.17a)–(3.17b) are feasible for some $Q(\cdot)$, $S(\cdot)$, $\hat{B}_K(\cdot)$, and $\hat{C}_K(\cdot)$. Then it is clear that the same quadruple together with $D_K(t) = 0$ and $\hat{A}_K(t) = -A^T(t)$ also satisfies (3.11a) since the anti-diagonal terms vanish and it reduces to (3.17a)–(3.17b). The conclusion is that, without loss of generality, we can restrict the search to matrix functions $Q(\cdot)$, $S(\cdot)$, $\hat{B}_K(\cdot)$, $\hat{C}_K(\cdot)$ satisfying (3.17a)–(3.17b) in place of (3.11a). ◊

Remark 3.2 (Controller Design) Assume now that the hypotheses of Theorem 3.2 are satisfied; in order to design the controller, the following steps have to be followed:

(i) Find $Q(\cdot)$, $S(\cdot)$, $N(\cdot)$, $\hat{A}_K(\cdot)$, $\hat{B}_K(\cdot)$, $\hat{C}_K(\cdot)$, and $D_K(\cdot)$ such that (3.11a)–(3.11c) are satisfied.
(ii) Let $M(t) = (I - S(t)Q(t))N^{-T}(t)$.
(iii) Obtain $A_K(\cdot)$, $B_K(\cdot)$, $C_K(\cdot)$, and $D_K(\cdot)$ by inverting (3.16b)–(3.16d).

Regarding step (i), according to Remark 3.1, one can let $\hat{A}_K(t) = -A^T(t)$ and $D_K = 0$ and restrict the search to $Q(\cdot)$, $S(\cdot)$, $N(\cdot)$, $\hat{B}_K(\cdot)$, and $\hat{C}_K(\cdot)$. ◊

Concerning the numerical implementation of the condition contained in the statement of Theorem 3.2 for the design of the output feedback controller, note that (3.11a) is a DLMI, condition (3.11b) is a time-varying LMI, while the initial condition (3.13c) *is not* an LMI, since it includes a quadratic term in the optimization variables $Q(t_0)$ and $N(t_0)$.

To render the problem computationally tractable, it is possible either to check it a posteriori or to replace, at the price of some conservatism, the term Δ_{11} by

$$Q(t_0)R^{1/2} + R^{1/2}Q(t_0) + N(t_0)R_K^{1/2} + R_K^{1/2}N(t_0)^T - 2I,$$

which is a lower bound of Δ_{11}. This fact can be easily derived from the following inequality:

$$\left(R^{1/2}Q(t_0) - I\right)^T \left(R^{1/2}Q(t_0) - I\right) + \left(R_K^{1/2}N(t_0)^T - I\right)^T \left(R_K^{1/2}N(t_0)^T - I\right) > 0.$$

In this way, condition (3.11c) is implied by

$$\begin{pmatrix} Q(t_0) & I \\ I & S(t_0) \end{pmatrix}$$
$$< \begin{pmatrix} Q(t_0)R^{1/2} + R^{1/2}Q(t_0) + N(t_0)R_K^{1/2} + R_K^{1/2}N(t_0)^T - 2I & Q(t_0)R \\ RQ(t_0) & R \end{pmatrix},$$

$$(3.18)$$

which is an LMI in the optimization variables.

The statement of Theorem 3.2 requires to find a nonsingular matrix-valued function $N(\cdot)$. To this end, we can enforce positive definiteness of $N(\cdot)$; indeed, the following result holds.

Lemma 3.2 *There exists a positive definite matrix-valued function $N(\cdot)$ defined for $t \in [t_0, t_0 + T]$, satisfying (3.11b) and (3.18), if and only if there exists a nonsingular matrix-valued sequence $\bar{N}(\cdot)$, defined for $t \in [t_0, t_0 + T]$, satisfying the same inequalities.* $\qquad\square$

Proof The necessity is obvious.

Now assume there exists a nonsingular matrix-valued function $\bar{N}(t)$, $t \in [t_0, t_0 + T]$, that satisfies (3.11b) and (3.18).

By continuity arguments, it is simple to recognize that there exists a symmetric positive definite matrix-valued function $N(\cdot)$ (set, for example, $N(t) = \gamma(t)I$ with $\sup_{t \in [t_0, t_0 + T]} \gamma(t)$ sufficiently small) such that

$$\begin{pmatrix} Q(t) & \Psi_{12}(t) & \Psi_{13}(t) & N(t)\Gamma_K^{1/2} \\ \Psi_{12}^T(t) & \Psi_{22}(t) & 0 & 0 \\ \Psi_{13}^T(t) & 0 & I & 0 \\ \Gamma_K^{1/2}N(t) & 0 & 0 & I \end{pmatrix} > 0, \quad t \in [t_0, t_0 + T],$$

$$\begin{pmatrix} Q(t_0) & I \\ I & S(t_0) \end{pmatrix}$$

$$< \begin{pmatrix} Q(t_0)R^{1/2} + R^{1/2}Q(t_0) + N(t_0)R_K^{1/2} + R_K^{1/2}N(t_0) - 2I & Q(t_0)R \\ RQ(t_0) & R \end{pmatrix}.$$

This completes the proof.   ◊

3.4 Finite-Time Stabilization of the Car Suspension System

In this section, we consider the active suspension system introduced in Sect. 2.5. By means of Theorem 2.1 we have shown that the system is finite-time stable with respect to $(t_0, T, R, \Gamma(\cdot))$ when we let $t_0 = 0$ s, $T = 1$ s, and

$$R = \text{diag}\left(\infty \quad \infty \quad \frac{1}{x_{3\,\text{max}}^2} \quad \infty \right), \tag{3.19a}$$

$$\Gamma(t) = \text{diag}\left(\frac{1}{(x_{1\,\text{max}}e^{-\frac{t}{\tau}})^2} \quad 0 \quad 0 \quad 0 \right), \tag{3.19b}$$

with $x_{1\,\text{max}} = 0.02$ m and $\tau = 0.4$ s, for all $|x_{3\,\text{max}}| < 0.049$ m (see Sect. 2.5.2), whereas the nonviolation of the constraints cannot be guaranteed for values of $|x_3(0)| \geq 0.049$. For instance, the simulation reported in Fig. 2.3 shows that, in the open-loop system, the suspension stroke exceeds the assigned threshold when $x_3(0) = 0.08$.

In this section, we exploit Theorem 3.1 to design a state-feedback controller that finite-time stabilizes the active suspension when

$$x_0 = \begin{pmatrix} 0 & 0 & 0.08 & 0 \end{pmatrix}^T.$$

This result can be achieved by requiring the closed-loop system to be finite-time stable wrt $(0, 1, R, \Gamma)$, where R is given by (3.19a) with $x_{3\,\text{max}} = 0.08$ m, and $\Gamma(\cdot)$ is given by (3.19b).

According to Sect. 2.4, in order to recast the DLMI problem (3.6a)–(3.6c) in terms of an LMI feasibility problem, the matrix-valued functions $Q(\cdot)$ and $L(\cdot)$ have been assumed to be piecewise linear. Therefore, the following structure is assigned to the matrix-valued function $Q(\cdot)$:

$$Q(t) = \Omega_j^0 + \Omega_j^*\big(t - (j-1)T_s\big), \quad t \in \big[(j-1)T_s, jT_s\big], \quad j = 1, \ldots, J+1,$$

where $J = \max\{j \in \mathbb{N} : j < T/T_s\}$, and Ω_j^0 and Ω_j^* are the optimization matrices. The time interval Ω has been divided into $J + 1$ subintervals, and the time derivatives of $Q(\cdot)$ have been considered constant in each subinterval. The same structure is used also for the matrix-valued function $L(\cdot)$.

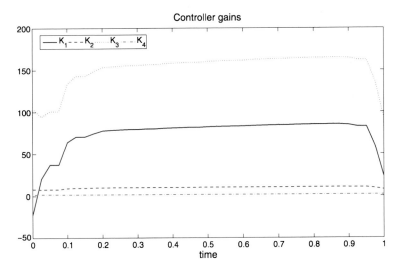

Fig. 3.1 Time-varying piecewise linear gains designed for the state feedback controller in Sect. 3.4

Exploiting optimization tools like the Matlab LMI Toolbox [48] and the YALMIP parser [70], it is easy to implement the conditions of Theorem 3.1 and to compute the optimization matrices defining $Q(\cdot)$ and $L(\cdot)$ and, then, the controller matrix-valued function $K(\cdot)$. The time-varying state-feedback controller

$$K(t) = \begin{pmatrix} K_1(t) & K_2(t) & K_3(t) & K_4(t) \end{pmatrix}$$

with the gains reported in Fig. 3.1 has been designed to render the closed-loop system finite-time stable wrt to $(0, 1, R, \Gamma(\cdot))$ with $x_{3\,max} = 0.08$ m. The effectiveness of the designed control system is confirmed by the simulation depicted in Figs. 3.2 and 3.3.

In order to moderate the control effort, the design optimization problem has been implemented along with additional pole placement LMI conditions (see [1, Sect. 3.4], [7]). The latter conditions allow the constraining of the poles of the closed-loop system within convex regions of the complex plane. We have exploited such additional constrains to set upper bounds on the convergence speeds and oscillations of the trajectories.

3.5 Summary

In this chapter, the finite-time stabilization problem has been discussed. Starting from the analysis result based on the DLMI condition, stated in (v) of Theorem 2.1, necessary and sufficient conditions for both state and output feedback finite-time stabilization have been provided. In the state feedback case, the condition is directly stated in the form of coupled DLMI/LMIs, while, in the output feedback case, the

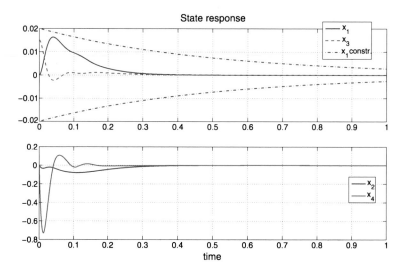

Fig. 3.2 State response of the state-feedback closed-loop system (2.20) with $u(t) = K(t)x(t)$ and the matrix function $K(t)$ shown in Fig. 3.1; the initial condition is equal to $(0\,0\,0.08\,0)$

Fig. 3.3 Control input for the state-feedback closed-loop system (2.20) with $u(t) = K(t)x(t)$ and the matrix function $K(t)$ shown in Fig. 3.1; the initial condition is equal to $(0\,0\,0.08\,0)$

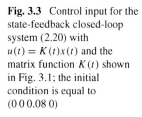

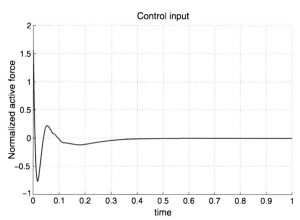

conditions are turned into DLMI/LMIs at the price of some conservatism. The application of the proposed technique has been illustrated with reference to the design of a state feedback controller for the car suspension system, whose model has been developed in Sect. 2.5.

The rest of the section will be devoted to illustrate some of the results, concerning the FTS design, which appeared in the literature before Theorem 3.2 was available. The first result regarding this issue, dealing with LTI systems and the state feedback stabilization, followed the guidelines of the approach described in Sect. 2.6 and was published in [6].

More precisely, consider the LTI system

$$\dot{x}(t) = Ax(t) + Bu(t), \quad x(0) = x_0, \tag{3.20}$$

and the state feedback controller of the form

$$u(t) = Kx(t), \tag{3.21}$$

where $K \in \mathbb{R}^{m \times n}$.

In the following problem, we refer to Definition 2.2.

Problem 3.3 (Finite-Time Stabilization Problem of CT-LTI Systems Via State Feedback) Given system (3.20) and the triplet (T, R, Γ), find a state feedback controller of the form (3.21) such that the closed-loop system given by the interconnection of (3.20) with (3.21) is finite-time stable, according to Definition 2.2, with respect to (T, R, Γ).   ◇

A sufficient condition for the solution of Problem 3.3, when $\Gamma = \rho R$, is given by the following theorem.

Theorem 3.3 [10] *Given system (3.20), Problem 3.3 with $\Gamma = \rho R$, $0 < \rho < 1$, is solvable if, letting $\tilde{Q} = R^{-\frac{1}{2}} Q R^{-\frac{1}{2}}$, there exist a nonnegative scalar α, a positive definite matrix $Q \in \mathbb{R}^{n \times n}$, and a matrix $L \in \mathbb{R}^{m \times n}$ such that*

$$A\tilde{Q} + \tilde{Q}A^T + BL + L^T B^T - \alpha\tilde{Q} < 0, \tag{3.22a}$$

$$\text{cond}(Q) < \frac{1}{\rho} e^{-\alpha T}. \tag{3.22b}$$

In this case, a controller which solves the finite-time stabilization problem is $K = L\tilde{Q}^{-1}$.   □

Concerning the output feedback stabilization problem, the technique proposed in [12] follows a two-step procedure. In the following, we recap the main points of the approach proposed in [12]. Consider the LTI system

$$\dot{x}(t) = Ax(t) + Bu(t), \quad x(0) = x_0, \tag{3.23a}$$

$$y(t) = Cx(t). \tag{3.23b}$$

First, with the aid of Theorem 3.3, a state feedback controller of the form $u = Kx$ is designed, if exists, to finite-time stabilize system (3.23a).

Now, the state of system (3.23a) can be estimated via a classical Luenberger observer of the form

$$\dot{\xi}(t) = A\xi(t) + Bu(t) + L\big(C\xi(t) - y(t)\big),$$

$$= A\xi(t) + Bu(t) + LC\big(\xi(t) - x(t)\big), \quad \xi(0) = 0. \tag{3.24}$$

Note that the observer may destroy the FTS attained by the state feedback controller due to the inaccuracy of the state estimate during the transients.

If we feedback the state estimate via the controller $u(t) = K\xi(t)$, we obtain the closed-loop state equations

$$\dot{x}(t) = Ax(t) + BK\xi(t), \quad x(0) = x_0, \tag{3.25a}$$

$$\dot{\xi}(t) = -LCx(t) + (A + BK + LC)\xi(t), \quad \xi(0) = 0. \tag{3.25b}$$

It is well known that by using the state transformation that brings the system into the state-estimation error base

$$\begin{pmatrix} x \\ e \end{pmatrix} = \begin{pmatrix} I & 0 \\ I & -I \end{pmatrix} \begin{pmatrix} x \\ \xi \end{pmatrix},$$

the system dynamics (3.25a)–(3.25b) can be rewritten as

$$\dot{x}(t) = (A + BK)x(t) - BKe(t), \quad x(0) = x_0, \tag{3.26}$$

with

$$\dot{e}(t) = (A + LC)e(t), \quad e(0) = x_0. \tag{3.27}$$

Therefore, the system state evolution is determined by the closed-loop matrix $(A + BK)$ and by the behavior of the exogenous input $e(\cdot)$. Obviously, for $e(t) = 0$, system (3.26) is finite-time stable, while the presence of a nonzero $e(t)$ could bring the state vector $x(t)$ outside the trajectory set.

In [12] a procedure to design an observer gain L in (3.24) such that the FTS property of the system $\dot{x} = (A + BK)x$ is not lost in presence of the estimation error is proposed; such procedure, which again leads to an LMI-based synthesis, exploits the concept of *finite-time boundedness* [5, 6].

Roughly speaking, a system is said to be finite-time bounded if, given a certain initial set, the state vector does not exit the trajectory set in presence of inputs belonging to a prespecified class. In recent years, the concept of finite-time boundedness has been replaced by that of *input–output finite-time stability* [17] (see Sect. 1.2.1), which is not dealt with in this book.

We end this section by observing that, following the guidelines of the proof of Theorem 3.2, a procedure for the design in one shot of the output feedback controller, in the FTS context pictured by Definition 2.2, can be given; the following results are given here for the first time.

First, note that the output feedback control problem corresponding to Problem 3.3 can be stated as follows. Given system (3.23a)–(3.23b), consider the dynamical controller

$$\dot{\xi}(t) = A_K\xi(t) + B_K y(t), \quad \xi(0) = 0, \tag{3.28a}$$

$$u(t) = C_K\xi(t) + D_K y(t). \tag{3.28b}$$

Problem 3.4 (Finite-Time Stabilization of CT-LTI System Via Output Feedback)
Given $T > 0$, two positive definite matrices R and R_K and two positive definite
matrices Γ and Γ_K with $\Gamma < R$ and $\Gamma_K < R_K$, find a dynamical output feedback
controller of the form (3.28a)–(3.28b) such that the closed-loop system obtained
by the connection of (3.23a)–(3.23b) and (3.28a)–(3.28b) is finite-time stable wrt
$(T, \mathrm{blockdiag}(R, R_k), \mathrm{blockdiag}(\Gamma, \Gamma_K))$.   ◇

The following result gives an answer to Problem 3.4 when, as usual, the trajectory
sets are proportional to the initial sets.

Theorem 3.4   *Given system (3.23a)–(3.23b), Problem 3.4 with $\Gamma = \rho R$ and $\Gamma_K =
\rho R_K$, $0 < \rho < 1$, is solvable if there exist a nonnegative scalar α, symmetric ma-
trices Q and S, a nonsingular matrix N, and matrices $\hat{A}_K$, $\hat{B}_K$, $\hat{C}_K$, and D_K such
that*

$$\begin{pmatrix} \Theta_{11} & \Theta_{12} \\ \Theta_{21} & \Theta_{22} \end{pmatrix} < 0, \tag{3.29a}$$

$$\begin{pmatrix} Q & I - \gamma Q & Q & N \\ I - \gamma Q & S - \gamma I & 0 & 0 \\ Q & 0 & \frac{1}{\gamma} I & 0 \\ N^T & 0 & 0 & \frac{1}{\gamma} I \end{pmatrix} > 0, \tag{3.29b}$$

$$\begin{pmatrix} Q - Q^2 - NN^T & I - Q \\ I - Q & S - I \end{pmatrix} < 0, \tag{3.29c}$$

where $\gamma = \rho e^{\alpha T}$, and

$\Theta_{11} := R^{1/2} A R^{1/2} Q + Q R^{-1/2} A^T R^{1/2} + R^{1/2} B \hat{C}_K + \hat{C}_K^T B^T R^{1/2} - \alpha Q,$

$\Theta_{12} := \Theta_{21}^T := R^{1/2} A R^{-1/2} + \hat{A}_K^T + R^{1/2} B D_K C R^{-1/2} - \alpha I,$

$\Theta_{22} := S R^{1/2} A R^{-1/2} + R^{-1/2} A^T R^{1/2} S + \hat{B}_K C R^{-1/2} + R^{-1/2} C^T \hat{B}_K^T - \alpha S.$ □

Proof The closed-loop output feedback system, derived by the interconnection of
systems (2.24) and (3.28a)–(3.28b), reads

$$\begin{pmatrix} \dot{x}(t) \\ \dot{\xi}(t) \end{pmatrix} = \begin{pmatrix} A + B D_K C & B C_K \\ B_K C & A_K \end{pmatrix} \begin{pmatrix} x \\ \xi \end{pmatrix} =: A_{\mathrm{of}} x_{\mathrm{of}}(t), \tag{3.31}$$

where x_{of} and A_{of} denote the state vector and the system matrix of the closed-
loop system, respectively. By applying Theorem 2.2 to system (3.31) we have that
a sufficient condition for the FTS of the closed-loop system is the existence of a

nonnegative scalar α, a positive definite matrix P_{of}, and matrices A_K, B_K, C_K, and D_K such that

$$A_{of}^T \tilde{P}_{of} + \tilde{P}_{of} A_{of} - \alpha \tilde{P}_{of} < 0, \tag{3.32a}$$

$$\mathrm{cond}(P_{of}) < \frac{1}{\rho} e^{-\alpha T}, \tag{3.32b}$$

where $\tilde{P}_{of} := R_{of}^{1/2} P_{of} R_{of}^{1/2}$ with $R_{of} := \mathrm{blockdiag}(R, R_K)$.

Note that the statement of the theorem and (3.29b) guarantee the existence of positive definite matrices Q and S such that

$$\begin{pmatrix} Q & I \\ I & S \end{pmatrix} > 0. \tag{3.33}$$

According to Lemma 3.1, condition (3.33) guarantees the existence of a symmetric matrix U and nonsingular matrices M and N such that

$$P_{of} = \begin{pmatrix} S & M \\ M^T & U \end{pmatrix}, \qquad Q_{of} := P_{of}^{-1} = \begin{pmatrix} Q & N \\ N^T & \star \end{pmatrix}, \tag{3.34}$$

$$\Pi_1 = \begin{pmatrix} Q & I \\ N^T & 0 \end{pmatrix}, \qquad \Pi_2 = \begin{pmatrix} I & S \\ 0 & M^T \end{pmatrix}, \tag{3.35}$$

where $\star$ denotes a "does not care" block.

Note that, by definition,

$$SQ + MN^T = I, \tag{3.36a}$$

$$M^T Q + U N^T = 0, \tag{3.36b}$$

$$P_{of} \Pi_1 = \Pi_2. \tag{3.36c}$$

The equivalence of (3.29a) and (3.32a) is derived by pre- and post-multiplying (3.32a) by $\Pi_1^T R_{of}^{-1/2}$ and $R_{of}^{-1/2} \Pi_1$, respectively, taking into account (3.36a)–(3.36c), and letting

$$\hat{B}_K := M R_K^{1/2} B_K + S R^{1/2} B D_K, \tag{3.37a}$$

$$\hat{C}_K := C_K R_K^{-1/2} N^T + D_K C R^{-1/2} Q, \tag{3.37b}$$

$$\hat{A}_K := M R_K^{1/2} A_K R_K^{-1/2} N^T + S R^{1/2} B C_K R_K^{-1/2} N^T,$$

$$\qquad + M R_K^{1/2} B_K C R^{-1/2} Q + S R^{1/2} (A + B D_K C) R^{-1/2} Q. \tag{3.37c}$$

Inequality (3.32b) can be guaranteed by imposing the conditions

$$\rho e^{\alpha T} I < \begin{pmatrix} S & M \\ M^T & U \end{pmatrix} < I, \tag{3.38}$$

which are LMIs in the variables Q, M, and U. However, such conditions are not useful in this form, since they would yield values of Q, M, and U that are not constrained to satisfy conditions (3.36a)–(3.36c). On the other hand, imposing such equality constraints would lead to a non-LMI problem. To overcome this limitation, let us pre- and post-multiply also both sides of (3.38) by Π_1^T and Π_1, respectively. By easy calculations, the left inequality is transformed (by applying the properties of Schur complements) into the LMI (3.29b). The right inequality, instead, becomes (3.29c). ◊

Note that the statement of Theorem 3.4 requires to find a nonsingular N; this can be obtained by adding a further LMI constraint requiring the positive definiteness of N.

When the hypotheses of Theorem 3.4 are satisfied, in order to design the controller, the following steps have to be followed:

(i) Find Q, S, N, $\hat{A}_K$, $\hat{B}_K$, $\hat{C}_K$, and D_K such that conditions (3.29a)–(3.29c) are satisfied;
(ii) Let $M = (I - SQ)N^{-T}$;
(iii) Obtain A_K, B_K, C_K by inverting (3.37a)–(3.37c). ◊

Note that conditions (3.29a) and (3.29b) are LMIs, whereas (3.29c) is not, due to the presence of the quadratic term in the upper-left block

$$Q - Q^2 - N N^T, \tag{3.39}$$

which cannot be linearized by applying the usual Schur complement machinery.

In order to obtain an LMI problem, we can replace (3.39) by

$$Q - N Q - Q N^T. \tag{3.40}$$

Indeed, by completion of squares we derive

$$0 < \left(Q + N^T \right)^T \left(Q + N^T \right)$$

$$= Q^2 + N Q + Q N + N N^T$$

$$\Rightarrow -Q^2 - N N^T < N Q + Q N^T.$$

Chapter 4
Robustness Issues

4.1 Introduction

In this chapter, we assume that the system under consideration is subject to uncertainties. The uncertain system topology dealt with is similar to the usual robustness setting considered in the context of the classical LS; indeed, we consider uncertainties modeled via the classical LFT form [45, 90]. However, the two situations are rather different; for instance, in the time-invariant context, the LS robustness is guaranteed if the uncertainties are such that the poles of the perturbed system matrix do not cross the imaginary axis; conversely, in the FTS case, there is no correlation between stability and location of poles on the complex plane; therefore, the situation is much more intricate.

To correctly frame our robustness results, the novel concept of QFTS is defined. Roughly speaking, a system is said to be quadratically finite-time stable if there exists a quadratic Lyapunov function that allows us to prove FTS of the given system for all admissible values of the uncertainties.

The first result of the chapter is a necessary and sufficient condition for QFTS, provided in terms of both DLMI- and DLE-based conditions, respectively. The analysis result represents the starting point for the design of state feedback and output feedback controllers that render the closed-loop system quadratically finite-time stable. The methodology is again tested versus the quarter car system model developed in Chap. 2.

4.2 QFTS of Uncertain Linear Systems

This chapter deals with the uncertain version of system (2.24), namely

$$\dot{x} = \big(A(t) + F(t)\Delta(t)E(t)\big)x(t), \tag{4.1}$$

where $x(t) \in \mathbb{R}^n$, $\Delta(\cdot)$ is a norm-bounded uncertainty matrix satisfying $\|\Delta(t)\| \leq 1$, and $F(\cdot)$ and $E(\cdot)$ are weighting matrices of suitable dimensions. All matrix-valued

F. Amato et al., *Finite-Time Stability and Control*,
Lecture Notes in Control and Information Sciences 453,
DOI 10.1007/978-1-4471-5664-2_4, © Springer-Verlag London 2014

Fig. 4.1 The uncertain
system (4.3a)–(4.3b) in LFT
form

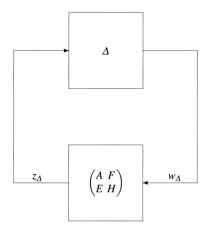

functions, unless otherwise specified, are assumed to be bounded and piecewise continuous in $[t_0, t_0 + T]$.

Note that system (4.1) is equivalent to the classical closed-loop system given by the *Linear Fractional Transformation* (LFT) between the linear system

$$\dot{x}(t) = A(t)x(t) + F(t)w_\Delta(t),$$

$$z_\Delta(t) = E(t)x(t),$$

and the uncertainty Δ, according to

$$w_\Delta(t) = \Delta(t)z_\Delta(t). \tag{4.2}$$

If we look at the LFT interpretation of norm-bounded uncertainties depicted in Fig. 4.1, rather than considering the uncertainty as perturbations of the system matrix, we can analyze the more general case in which the nominal system is proper, i.e., it has a nonzero feedthrough matrix (see [1, Chap. 3]). In this case, the uncertain system is described by the closed-loop connection between the system

$$\dot{x}(t) = A(t)x(t) + F(t)w_\Delta(t), \tag{4.3a}$$

$$z_\Delta(t) = E(t)x(t) + H(t)w_\Delta(t), \tag{4.3b}$$

and (4.2) or, equivalently,

$$\dot{x}(t) = \big(A(t) + \Delta A(t)\big)x(t), \tag{4.4a}$$

where

$$\Delta A(t) = F(t)\Delta(t)\big(I - H(t)\Delta(t)\big)^{-1}E(t)$$

$$= F(t)\big(I - \Delta(t)H(t)\big)^{-1}\Delta(t)E(t). \tag{4.4b}$$

Remark 4.1 Note that (4.4a)–(4.4b) is well posed if $I - H(t)\Delta(t)$ is invertible for all t and Δ with $\|\Delta(t)\| \leq 1$, $t \in [t_0, t_0 + T]$; this is equivalent to the requirement that $\|H(t)\| < 1$, $t \in [t_0, t_0 + T]$; therefore, this assumption will be implicitly done in the following. On the other hand, the conditions for FTS we will find later automatically guarantee the satisfaction of such an assumption. ◇

The following definition of *Robust Finite-Time Stability* (RFTS) is given for system (4.4a)–(4.4b).

Definition 4.1 (RFTS of CT-LTV Systems) Given an initial time t_0, a positive scalar T, a positive definite matrix R, and a positive definite matrix-valued function $\Gamma(\cdot)$, defined over $[t_0, t_0 + T]$, such that $\Gamma(t_0) < R$, system (4.4a)–(4.4b) is said to be *robust finite-time stable* (RFTS) wrt $(t_0, T, R, \Gamma(\cdot))$ if

$$x_0^T R x_0 \leq 1 \Rightarrow x(t)^T \Gamma(t) x(t) < 1, \quad t \in [t_0, t_0 + T],$$

for all $\Delta(\cdot)$ with $\|\Delta(t)\| \leq 1$, $t \in [t_0, t_0 + T]$. ◇

We now introduce the definition of QFTS for system (4.4a)–(4.4b). QFTS implies FTS for any admissible uncertainty realization $\Delta(\cdot)$, and hence it implies RFTS. It should be noticed that the converse in not necessarily true, and hence QFTS is a *stronger* property when compared with RFTS. The *quadratic extension* of FTS, when dealing with uncertain systems, follows from what has been proposed in the literature in the case of LS (see, for example, [1, 3, 31, 43])

Definition 4.2 (QFTS) Given an initial time t_0, a positive scalar T, a positive definite matrix R, and a positive definite matrix-valued function $\Gamma(\cdot)$, defined over $[t_0, t_0 + T]$, such that $\Gamma(t_0) < R$, system (4.4a)–(4.4b) is said to be quadratic finite-time stable with respect to $(t_0, T, R, \Gamma(\cdot))$ if and only if there exists a positive definite matrix-valued function $P(\cdot)$ that satisfies the following DLMI with terminal and initial conditions:

$$\dot{P}(t) + \big(A(t) + \Delta A(t)\big)^T P(t) + P(t)\big(A(t) + \Delta A(t)\big) < 0, \quad t \in [t_0, t_0 + T],$$
(4.5a)

$$P(t) > \Gamma(t), \quad t \in [t_0, t_0 + T],$$
(4.5b)

$$P(t_0) < R,$$
(4.5c)

for any admissible uncertainty realization $\Delta(\cdot)$. ◇

The following lemma readily follows from Theorem 2.1 and from the definition of QFTS.

Lemma 4.1 *If system (4.4a)–(4.4b) is QFTS wrt $(t_0, T, R, \Gamma(\cdot))$, then it is RFTS wrt $(t_0, T, R, \Gamma(\cdot))$.* □

It is worth noticing that finding a solution to (4.5a)–(4.5c) is practically impossible since (4.5a) is an infinite-dimensional DLMI, that is, all possible realizations of $\Delta(\cdot)$ should be considered. The next theorem provides a necessary and sufficient condition based on a single DLMI that enables us to check QFTS of the uncertain system (4.4a)–(4.4b) in a numerically efficient way.

The sufficiency part of the next theorem (with the function $\lambda(\cdot)$ set to 1) has been published in [23]; the necessary part of the theorem is detailed here for the first time.

Theorem 4.1 *System* (4.4a)–(4.4b) *is QFTS wrt* $(t_0, T, R, \Gamma(\cdot))$ *if and only if there exist a symmetric piecewise continuously differentiable matrix-valued function* $P(\cdot)$ *and a function* $\lambda(\cdot) > 0$, *such that, for* $t \in [t_0, t_0 + T]$,

$$\begin{pmatrix} \dot{P}(t) + A^T(t)P(t) + P(t)A(t) + \lambda(t)E^T(t)E(t) & P(t)F(t) + \lambda(t)E^T(t)H(t) \\ F^T(t)P(t) + \lambda(t)H^T(t)E(t) & -\lambda(t)(I - H^T(t)H(t)) \end{pmatrix}$$

$$< 0, \tag{4.6a}$$

$$P(t) > \Gamma(t), \tag{4.6b}$$

$$P(t_0) < R. \tag{4.6c}$$

$\square$

Proof Hereafter we will omit the time-dependence for the sake of brevity. First of all, note that (4.6a) implies $I - H^T H > 0$; thus, we have

$$\|H\Delta\| \le \|H\|\|\Delta\|$$

$$\le \|H\| < 1;$$

this guarantees that $I - H\Delta$ is nonsingular for all Δ such that $\|\Delta\| \le 1$.

Note that, by replacing ΔA with the expression in (4.4b), condition (4.5a) reads

$$\dot{P} + \left(A + F(I - \Delta H)^{-1}\Delta E\right)^T P + P\left(A + F(I - \Delta H)^{-1}\Delta E\right) < 0,$$

$$t \in [t_0, t_0 + T], \tag{4.7}$$

subject to $\|\Delta(t)\| \le 1, t \in [t_0, t_0 + T]$.

Now condition (4.7) can be equivalently rewritten

$$x^T\left(\dot{P} + A^T P + PA\right)x + 2x^T PFv$$

$$= \begin{pmatrix} x^T & v^T \end{pmatrix}^T \begin{pmatrix} \dot{P} + A^T P + PA & PF \\ F^T P & 0 \end{pmatrix} \begin{pmatrix} x \\ v \end{pmatrix} < 0 \tag{4.8}$$

for all $x \in \mathbb{R}^n$, $x \ne 0$, $v \in S_x$, where

$$S_x := \left\{v \in \mathbb{R}^n : v = \Delta(Hv + Ex), \|\Delta\| \le 1\right\}$$

$$= \left\{v \in \mathbb{R}^n : v^T v \le (Hv + Ex)^T(Hv + Ex)\right\}. \tag{4.9}$$

By applying the S-procedure [38, p. 24], it can be shown that (4.8)–(4.9) are equivalent to the existence of a scalar function $\lambda(\cdot)$ such that (4.6a) holds. ◊

4.3 Quadratic Finite-Time Stabilization of Uncertain Linear Systems

In this section, the problems of quadratic finite-time stabilization via state and output feedback are first introduced. Afterwards, sufficient conditions to solve these problems are provided.

Problem 4.1 (Quadratic Finite-Time Stabilization Via State Feedback) Let us consider the uncertain time-varying linear system

$$\dot{x}(t) = \big(A(t) + \Delta A(t)\big)x(t) + \big(B(t) + \Delta B(t)\big)u(t)$$

with

$$\big(\Delta A(t) \quad \Delta B(t)\big) = F(t)\Delta(t)\big(I - H(t)\Delta(t)\big)^{-1}\big(E_1(t) \quad E_2(t)\big),$$

where $u(t) \in \mathbb{R}^m$ is the control input, and $F(\cdot)$, $E_1(\cdot)$, and $E_2(\cdot)$ are weighting matrix-valued functions of suitable dimensions. The uncertain system can be transformed into the LFT form

$$\dot{x}(t) = A(t)x(t) + F(t)w_\Delta(t) + B(t)u(t), \tag{4.10a}$$

$$z_\Delta(t) = E_1(t)x(t) + H(t)w_\Delta(t) + E_2(t)u(t), \tag{4.10b}$$

$$w_\Delta(t) = \Delta(t)z_\Delta(t). \tag{4.10c}$$

Then given an initial time t_0, a positive scalar T, a positive definite matrix R, and a positive definite matrix-valued function $\Gamma(\cdot)$, defined over $[t_0, t_0 + T]$, such that $\Gamma(t_0) < R$, find a state feedback controller of the form

$$u(t) = K(t)x(t) \tag{4.11}$$

such that the closed-loop system obtained by the connection of (4.10a)–(4.10c) and (4.11), namely

$$\dot{x}(t) = \big(A(t) + B(t)K(t)\big)x(t) + F(t)w_\Delta(t), \tag{4.12a}$$

$$z_\Delta(t) = \big(E_1(t) + E_2(t)K(t)\big)x(t) + H(t)w_\Delta(t), \tag{4.12b}$$

$$w_\Delta(t) = \Delta(t)z_\Delta(t), \tag{4.12c}$$

is quadratically finite-time stable with respect to $(t_0, T, R, \Gamma(\cdot))$. ◊

Problem 4.2 (Quadratic Finite-Time Stabilization Via Output Feedback) Consider the uncertain time-varying linear system

$$\dot{x}(t) = \big(A(t) + \Delta A(t)\big)x(t) + \big(B(t) + \Delta B(t)\big)u(t), \qquad (4.13a)$$

$$y(t) = \big(C(t) + \Delta C(t)\big)x(t) + \big(D(t) + \Delta D(t)\big)u(t), \qquad (4.13b)$$

with

$$\begin{pmatrix} \Delta A(t) & \Delta B(t) \\ \Delta C(t) & \Delta D(t) \end{pmatrix} = \begin{pmatrix} F_1(t) \\ F_2(t) \end{pmatrix} \Delta(t)\big(I - H(t)\Delta(t)\big)^{-1} \big(E_1(t) \quad E_2(t)\big),$$

where $y(t) \in \mathbb{R}^r$ is the output, and F_i and E_i, $i = 1, 2$, are weighting matrices of suitable dimensions. Adopting the LFT form, the system can be rewritten as

$$\dot{x}(t) = A(t)x(t) + F_1(t)w_\Delta(t) + B(t)u(t), \qquad (4.14a)$$

$$z_\Delta(t) = E_1(t)x(t) + H(t)w_\Delta(t) + E_2(t)u(t), \qquad (4.14b)$$

$$y(t) = C(t)x(t) + F_2(t)w_\Delta(t) + D(t)u(t), \qquad (4.14c)$$

$$w_\Delta(t) = \Delta(t)z_\Delta(t). \qquad (4.14d)$$

In the following, we assume, for the sake of simplicity, but without loss of generality, that $D(\cdot) = 0$. Then, given an initial time t_0, a positive scalar T, two positive definite matrices R and R_K, and two positive definite symmetric matrix functions $\Gamma(\cdot)$ and $\Gamma_K(\cdot)$, defined over $[t_0, t_0 + T]$, such that $\Gamma(t_0) < R$ and $\Gamma_K(t_0) < R_K$, find a dynamical output feedback controller of the form

$$\dot{x}_c(t) = A_K(t)x_c(t) + B_K(t)y(t), \qquad (4.15a)$$

$$u(t) = C_K(t)x_c(t) + D_K(t)y(t), \qquad (4.15b)$$

where $x_c(t) \in \mathbb{R}^n$, such that the closed-loop system obtained by the connection of (4.14a)–(4.14d) and (4.15a)–(4.15b) is quadratically finite-time stable with respect to

$$\big(t_0, T, \text{blockdiag}(R, R_K), \text{blockdiag}\big(\Gamma(\cdot), \Gamma_K(\cdot)\big)\big). \qquad \Diamond$$

In order to end up with DLMIs also in the design case, we introduce, at the price of some conservativeness, the following corollary of Theorem 4.1, where $\lambda(\cdot)$ is chosen to be a *constant*.

Corollary 4.1 *System* (4.4a)–(4.4b) *is well posed and QFTS with respect to* $(t_0, T, R, \Gamma(\cdot))$ *if the inequalities*

$$\begin{pmatrix} \dot{P}(t) + A(t)^T P(t) + P(t) A(t) & P(t) F(t) & E^T(t) \\ F^T(t) P(t) & -I & H^T(t) \\ E(t) & H(t) & -I \end{pmatrix} < 0, \quad t \in [t_0, t_0 + T],$$

$$\tag{4.16a}$$

$$P(t) > \Gamma(t), \quad t \in [t_0, t_0 + T], \tag{4.16b}$$

$$P(t_0) < R, \tag{4.16c}$$

admit a piecewise continuously differentiable symmetric solution $P(\cdot)$.

Proof Let us consider conditions (4.6a)–(4.6c) of Theorem 4.1 in the case of a constant $\lambda > 0$. Under this assumption, dividing all terms of (4.6a) by λ and rescaling the matrix function $P(\cdot)$ by the factor λ, we obtain the equivalent inequality

$$\begin{pmatrix} \dot{P}(t) + A(t)^T P(t) + P(t) A(t) + E^T(t) E(t) & P(t) F(t) + E^T(t) H(t) \\ F^T(t) P(t) + H^T(t) E(t) & -(I - H^T(t) H(t)) \end{pmatrix} < 0. \tag{4.17}$$

By applying the properties of Schur complements, the latter inequality can be equivalently rewritten as (4.16a).   ◊

4.3.1 The State Feedback Case

The following result provides a solution to Problem 4.1.

Theorem 4.2 [23]   *Problem* 4.1 *admits a solution if there exist a piecewise continuously differentiable symmetric matrix-valued function* $Q(\cdot)$ *and a matrix-valued function* $L(\cdot)$ *such that*[1]

$$\begin{pmatrix} -\dot{Q} + Q A^T + A Q + L^T B^T + B L & F & Q E_1^T + L^T E_2^T \\ \cdots & -I & H^T \\ \cdots & \cdots & -I \end{pmatrix} < 0,$$

$$t \in [t_0, t_0 + T], \tag{4.18a}$$

$$Q(t) < \Gamma^{-1}(t), \quad t \in [t_0, t_0 + T], \tag{4.18b}$$

$$Q(t_0) > R^{-1}. \tag{4.18c}$$

In this case, a controller gain that solves Problem 4.1 *is* $K(t) = L(t) Q^{-1}(t)$.   □

---

[1]For the sake of brevity, in the DLMI, we will omit the time argument and the lower triangular entries of the symmetric matrices and replace them with dots.

Proof By applying Corollary 4.1 to the closed-loop system (4.12a)–(4.12c), it follows that Problem 4.1 admits a solution if there exist a piecewise continuously differentiable symmetric matrix-valued function $P(\cdot)$ and a matrix-valued function $K(\cdot)$ such that

$$\begin{pmatrix} \dot{P} + (A + BK)^T P + P(A + BK) & PF & (E_1 + E_2K)^T \\ \cdots & -I & H^T \\ \cdots & \cdots & -I \end{pmatrix} < 0,$$

$$t \in [t_0, t_0 + T], \tag{4.19a}$$

$$P(t) > \Gamma(t), \quad t \in [t_0, t_0 + T], \tag{4.19b}$$

$$P(t_0) < R. \tag{4.19c}$$

Now let $Q(t) = P^{-1}(t)$ and pre- and post-multiply (4.19a) by blockdiag$(Q(t), I, I)$. Condition (4.18a) is obtained noticing that

$$\dot{Q}(t) = -Q(t)\dot{P}(t)Q(t),$$

and letting $L(t) = K(t)Q(t)$ according to [51]. Conditions (4.18b) and (4.18c) are readily derived from (4.19b) and (4.19c), respectively, by inverting both sides of the inequalities.   ◊

4.3.2 The Output Feedback Case

The following is the main result of the section.

Theorem 4.3 [23]   *Problem* 4.2 *is solvable if there exist piecewise continuously differentiable symmetric matrix-valued functions* $Q(\cdot)$ *and* $S(\cdot)$*, a nonsingular matrix-valued function* $N(\cdot)$*, and matrix-valued functions* $\hat{A}_K(\cdot)$*,* $\hat{B}_K(\cdot)$*,* $\hat{C}_K(\cdot)$*, and* $D_K(\cdot)$ *such that*

$$\begin{pmatrix} \Theta_{11} & A + \hat{A}_K^T + BD_KC & F_1 + BD_K F_2 & QE_1^T + \hat{C}_K^T E_2^T \\ \cdots & \Theta_{22} & SF_1 + \hat{B}_K F_2 & E_1^T + C^T D_K^T E_2^T \\ \cdots & \cdots & -I & H^T + F_2^T D_K^T E_2^T \\ \cdots & \cdots & \cdots & -I \end{pmatrix} < 0,$$

$$t \in [t_0, t_0 + T], \tag{4.20a}$$

$$\begin{pmatrix} Q & I - Q\Gamma & Q & N \\ \cdots & S - \Gamma & 0 & 0 \\ \cdots & \cdots & \Gamma^{-1} & 0 \\ \cdots & \cdots & \cdots & \Gamma_K^{-1} \end{pmatrix} > 0, \quad t \in [t_0, t_0 + T], \tag{4.20b}$$

Fig. 4.2 LFT representation
of the closed-loop system
(4.14a)–(4.14d) and
(4.15a)–(4.15b)

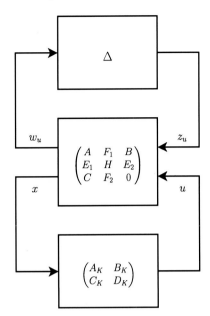

$$\begin{pmatrix} Q(t_0) & I \\ I & S(t_0) \end{pmatrix} < \begin{pmatrix} \Psi_{11} & Q(t_0)R \\ RQ(t_0) & R \end{pmatrix},$$ (4.20c)

where

$$\Theta_{11} = -\dot{Q} + AQ + QA^T + B\hat{C}_K + \hat{C}_K^T B^T,$$

$$\Theta_{22} = \dot{S} + SA + A^T S + \hat{B}_K C + C^T \hat{B}_K^T,$$

$$\Psi_{11} = Q(t_0)RQ(t_0) + N(t_0)R_K N^T(t_0).$$ □

Proof The closed-loop connection between system (4.14a)–(4.14d) and (4.15a)–(4.15b) is depicted in Fig. 4.2. From this figure it is clear that the closed loop is itself an unforced system subject to norm-bounded uncertainties and described by the equations

$$\dot{x}_{CL} = A_{CL}(t)x_{CL}(t) + F_{CL}(t)w_u(t),$$ (4.21a)

$$z_u(t) = E_{CL}(t)x_{CL}(t) + H_{CL}(t)w_u(t),$$ (4.21b)

$$w_u(t) = \Delta(t)z_u(t),$$ (4.21c)

where $x_{CL(t)} = (x^T(t)\ x_c^T(t))^T$, and

$$A_{CL}(t) = \begin{pmatrix} A(t) + B(t)D_K(t)C(t) & B(t)C_K(t) \\ B_K(t)C(t) & A_K(t) \end{pmatrix},$$

$$F_{\mathrm{CL}}(t) = \begin{pmatrix} F_1(t) + B(t)D_K(t)F_2(t) \\ B_K(t)F_2(t) \end{pmatrix},$$

$$E_{\mathrm{CL}}(t) = \begin{pmatrix} E_1(t) + E_2(t)D_K(t)C(t) & E_2(t)C_K(t) \end{pmatrix},$$

$$H_{\mathrm{CL}}(t) = H(t) + E_2(t)D_K(t)F_2(t).$$

Define $R_{\mathrm{CL}} = \mathrm{blockdiag}(R, R_K)$ and $\Gamma_{\mathrm{CL}}(t) = \mathrm{blockdiag}(\Gamma(t), \Gamma_K(t))$; according to Corollary 4.1, system (4.21a)–(4.21c) is RFTS wrt $(T, R_{\mathrm{CL}}, \Gamma_{\mathrm{CL}}(\cdot))$ if there exists a piecewise continuously differentiable symmetric matrix-valued function $P(\cdot)$ such that

$$\begin{pmatrix} \dot{P}(t) + A_{\mathrm{CL}}^T(t)P(t) + P(t)A_{\mathrm{CL}}(t) & P(t)F_{\mathrm{CL}}(t) & E_{\mathrm{CL}}^T(t) \\ F_{\mathrm{CL}}^T(t)P(t) & -I & H_{\mathrm{CL}}^T(t) \\ E_{\mathrm{CL}}(t) & H_{\mathrm{CL}}(t) & -I \end{pmatrix} < 0, \quad t \in [0, T],$$

$$\tag{4.22a}$$

$$P(t) \ge \Gamma_{\mathrm{CL}}(t), \quad t \in [0, T], \tag{4.22b}$$

$$P(t_0) < R_{\mathrm{CL}}. \tag{4.22c}$$

Now let us define the matrix functions $P(\cdot)$, $\Pi_1(\cdot)$, and $\Pi_2(\cdot)$ as in (3.14a)–(3.14b); by pre- and post-multiplying (4.22a)–(4.22c) by $\mathrm{blockdiag}(\Pi_1^T(t), I, I)$ and $\mathrm{blockdiag}(\Pi_1(t), I, I)$, respectively, taking into account (3.15a)–(3.15c) and Lemma 3.1, the proof follows once we define $\hat{A}_K$, $\hat{B}_K$, and $\hat{C}_K$ as in (3.16a)–(3.16d). $\Diamond$

Concerning the controller design, the procedure described in Remark (3.2) still holds in the uncertain case.

Note that (4.20a) is a DLMI, while (4.20b) is a time-dependent LMI; concerning the initial condition (4.20c), which exhibits a quadratic term in the optimization variables $Q(t_0)$ and $N(t_0)$, it is possible either to check it a posteriori or to replace the term Ψ_{11} by

$$\bar{\Psi}_{11} = Q(t_0)R^{1/2} + R^{1/2}Q(t_0) + N(t_0)R_K^{1/2} + R_K^{1/2}N(t_0)^T - 2I.$$

Indeed,

$$\Psi_{11} - \bar{\Psi}_{11} = \left(R^{1/2}Q(t_0) - I\right)^T \left(R^{1/2}Q(t_0) - I\right)$$

$$+ \left(R_K^{1/2}N(t_0)^T - I\right)^T \left(R_K^{1/2}N(t_0)^T - I\right) > 0,$$

and therefore $\Psi_{11} > \bar{\Psi}_{11}$.

4.4 Finite-Time Stabilization of a Car Suspension System Affected by Uncertainty

In this section, the uncertain model (4.3a)–(4.3b) is exploited to deal with the active suspension system considered in Sects. 2.5 and 3.4 in the presence of an uncertain parameter M_u.

In particular, the uncertainty on the parameter M_u can be modeled by choosing

$$
F = \begin{pmatrix} 0 \\ 0 \\ 0 \\ 1 \end{pmatrix}, \qquad E = \left(0.2\frac{K_s}{M_u} \quad 0.2\frac{B_s}{M_u} \quad -0.2\frac{K_u}{M_u} \quad -0.2\frac{B_s}{M_u} \right),
$$

$$
H = 0, \qquad \Delta \in [-1, 1].
$$

which amounts to taking into account uncertainties up to 20 % on the elements on the last row of the A matrix.

Similarly to what has been done in Sect. 2.5.2, condition (iii) of Theorem 2.1 can be exploited to show that the nominal system (without uncertainty) is not finite-time stable when we let

$$
t_0 = 0 \text{ s,}
$$

$$
T = 1 \text{ s,}
$$

$$
R = \mathrm{diag}\left(\infty \quad \infty \quad \frac{1}{x_{3\,\mathrm{max}}^2} \quad \infty \right),
$$

$$
\Gamma(t) = \mathrm{diag}\left(\frac{1}{(x_{1\,\mathrm{max}}e^{-\frac{t}{\tau}})^2} \quad 0 \quad 0 \quad 0 \right),
$$

with $|x_{3\,\mathrm{max}}| \geq 0.0068$ m, $x_{1\,\mathrm{max}} = 0.02$ m, and $\tau = 0.4$ s.

Coherently, Theorem 4.1 shows that the uncertain system is not quadratically finite-time stable wrt $(0, 1, R, \Gamma(\cdot))$. Figure 4.3 shows the time evolution of the weighted norm $x^T(t)\Gamma(t)x(t)$ when the active suspension system evolves from the initial condition $x_0 = (0, 0, 0.015, 0)^T$, for several values of the uncertainty Δ in the range $[-1, 1]$.

In order to design a state feedback controller that guarantees that the uncertain closed-loop system is quadratically finite-time stable wrt $(0, 1, R, \Gamma(\cdot))$, it is possible to exploit the DLMI conditions of Theorem 4.2. As shown in Sect. 3.4, these conditions can be transformed into LMIs by assigning a piecewise affine structure to the involved matrix-valued functions. Thus, by solving an LMI feasibility problem, we can find the state feedback control law conferring the desired QFTS property.

The optimization problem has been solved for different values of $x_{3\,\mathrm{max}}$: the maximum value for which it has been possible to find a solution to the QFTS problem equals 0.039 m. The validity of the designed controller is confirmed by the simulation of the state response shown in Fig. 4.4. Figures 4.5 and 4.6 report the corresponding time-varying controller matrix and the control signal, respectively.

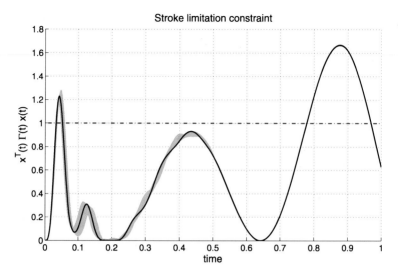

Fig. 4.3 Time history of the squared weighted norm of the state of the uncertain active suspension system; free evolution from the initial condition $(0\,0\,0.08\,0)$ for several values of $\Delta \in [-1, 1]$ (*the thick curve* corresponds to the nominal system, that is, $\Delta = 0$)

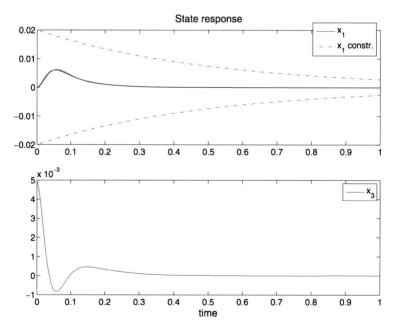

Fig. 4.4 State response of the closed-loop uncertain active suspension system: evolution from the initial condition $(0\,0\,0.039\,0)$ for several values of $\Delta \in [-1, 1]$ (*the thick curve* corresponds to the evolution of the nominal system $\Delta = 0$)

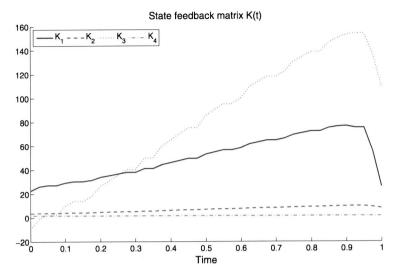

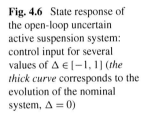

Fig. 4.5 State response of the open-loop uncertain active suspension system: time-varying controller matrix

Fig. 4.6 State response of
the open-loop uncertain
active suspension system:
control input for several
values of $\Delta \in [-1, 1]$ (*the
thick curve* corresponds to the
evolution of the nominal
system, $\Delta = 0$)

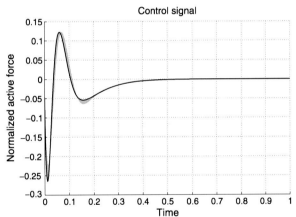

4.5 Summary

In this chapter, the FTS problem has been framed in the context of uncertain systems. The main result, appearing here for the first time, is a necessary and sufficient condition for QFTS of linear systems subject to norm-bounded uncertainties.

In the former paper [23], it was proven that the conditions stated in Theorem 4.1 guarantee RFTS in the presence of norm-bounded uncertainties. Theorem 4.1 shows that such conditions are actually also necessary for robust finite-stabilization when quadratic Lyapunov functions are used. Clearly, as in the LS case, QFTS implies RFTS, but the converse is not true since the class of quadratic Lyapunov functions is not universal (while polyhedral or piecewise quadratic Lyapunov functions are, see Chap. 6).

As usual, the analysis result represents the basis for the development of the design procedures. However, in order to end up with DLMIs both in the state and the output feedback case, the optimization scalar function λ is chosen constant (and, without loss of generality, unitary). This renders the synthesis conditions of Theorems 4.2 and 4.3 only sufficient for quadratic finite-time stabilization.

The technique is finally applied to design a robust state feedback controller for the car suspension system developed in the previous chapters.

Chapter 5
FTS of Discrete-Time Linear Systems

5.1 Introduction

In this chapter, we consider DT-LTV systems. Our main analysis theorem guarantees FTS *if and only if* either a certain inequality involving the state transition matrix is satisfied, or a symmetric matrix-valued sequence solving a certain DLE exists, or a symmetric matrix-valued sequence solving a certain DLMI exists. Such conditions can be considered the discrete-time counterpart of the conditions for FTS of CT-LTV systems stated in Theorem 2.1.

In principle, differently from the CT-LTV case, the condition involving the state transition matrix can be used to test the FTS of a given DT-LTV system; however, it is not useful to solve the synthesis problem. As done in Chap. 2, in the continuous-time context, the solution of the design problem will be obtained with the aid of the DLMI-based condition.

In particular, we shall solve both the state and output feedback finite-time stabilization problems. The state feedback solution is stated in the form of a necessary and sufficient condition subject to a coupled DLMI/LMIs condition; in the output feedback case, the condition is again necessary and sufficient; however, its implementation requires to modify, at the price of some conservativeness, the DLMI initial condition, which contains quadratic terms in the optimization variables, so to get a further LMI condition.

The chapter is ended by a numerical example that illustrates the theory developed.

5.2 Problem Statement

In this chapter, we consider the following DT-LTV system:

$$x(k+1) = A(k)x(k) + B(k)u(k), \qquad (5.1a)$$

$$y(k) = C(k)x(k), \qquad (5.1b)$$

where $A(k)$, $B(k)$, and $C(k)$ take values in $\mathbb{R}^{n \times n}$, $\mathbb{R}^{n \times m}$, and $\mathbb{R}^{p \times n}$, respectively.

F. Amato et al., *Finite-Time Stability and Control*,
Lecture Notes in Control and Information Sciences 453,
DOI 10.1007/978-1-4471-5664-2_5, © Springer-Verlag London 2014

In the following, we restrict Definition 1.1 to the case of autonomous DT-LTV systems, and we assume that both the initial set and the trajectory set are ellipsoids. An alternative possibility is that of considering polyhedral sets, as done, for instance, in [14, 49] (see also Chap. 6).

Definition 5.1 (FTS of DT-LTV Systems with Time-Varying Ellipsoidal Domains) Given an initial time k_0, a positive scalar N, a positive definite matrix R, and a positive definite matrix-valued sequence $\Gamma(\cdot)$, defined over $\{k_0, \ldots, k_0 + N\}$, such that $\Gamma(k_0) < R$, the DT-LTV system

$$x(k + 1) = A(k)x(k), \quad x(k_0) = x_0, \tag{5.2}$$

is said to be finite-time stable with respect to $(k_0, N, R, \Gamma(\cdot))$ if

$$x_0^T R x_0 \le 1 \Rightarrow x^T(k)\Gamma(k)x(k) < 1, \quad k \in \{k_0, \ldots, k_0 + N\}. \tag{5.3}$$
$\Diamond$

The rest of the section is devoted to define the finite-time stabilization problems in the discrete-time-context. First, we consider the state feedback stabilization; given the system

$$x(k + 1) = A(k)x(k) + B(k)u(k), \tag{5.4}$$

we consider the linear time-varying state feedback controller

$$u(k) = G(k)x(k), \tag{5.5}$$

where G takes values in $\mathbb{R}^{m \times n}$. One of the goals of this section is to find some necessary and sufficient conditions that guarantee that the state of the system given by the interconnection of system (5.4) with the controller (5.5) is *stable over a finite time interval*.

Problem 5.1 (Finite-Time Stabilization of DT-LTV Systems Via State Feeedback) Given an initial time k_0, an integer N, a positive definite matrix R, and a positive definite matrix-sequence $\Gamma(\cdot)$, find a state feedback controller (5.5) such that the closed-loop system obtained by the connection of (5.4) and (5.5), namely

$$x(k + 1) = \big(A(k) + B(k)G(k)\big)x(k), \tag{5.6}$$

is finite-time stable with respect to $(k_0, N, R, \Gamma(\cdot))$. $\Diamond$

Then with respect to system (5.1a)–(5.1b), we consider the following dynamical output feedback controller:

$$x_c(k + 1) = A_K(k)x_c(k) + B_K(k)y(k), \tag{5.7a}$$
$$u(k) = C_K(k)x_c(k) + D_K(k)y(k), \tag{5.7b}$$

where the controller state vector $x_c(k)$ has the same dimension of $x(k)$.

Problem 5.2 (Finite-Time Stabilization of DT-LTV Systems Via Output Feedback)
Given an initial time k_0, an integer N, two positive definite matrices R and R_K,
and two positive definite matrix-sequences $\Gamma(\cdot)$ and $\Gamma_K(\cdot)$, find an output feedback
controller of the form (5.7a)–(5.7b) such that the closed-loop system obtained by
the connection of (5.1a)–(5.1b) and (5.7a)–(5.7b) is finite-time stable with respect
to

$$\left(k_0, N, \operatorname{diag}(R, R_K), \operatorname{diag}\left(\Gamma(\cdot), \Gamma_K(\cdot)\right)\right). \qquad \Diamond$$

5.3 Analysis Conditions for FTS

The following theorem is fundamental for the subsequent development.

Theorem 5.1 (Necessary and Sufficient Conditions for FTS) *The following statements are equivalent*:

(i) *System (5.2) is finite-time stable with respect to $(k_0, N, R, \Gamma(\cdot))$.*
(ii) $\Phi^T(k, k_0)\Gamma(k)\Phi(k, k_0) < R$ *for all* $k \in \{k_0 + 1, \ldots, k_0 + N\}$, *where* $\Phi(\cdot, \cdot)$
denotes the state transition matrix of system (5.2).
(iii) *There exists a symmetric matrix-valued sequence* $P(\cdot) : k \in \{k_0, \ldots,$
$k_0 + N\} \mapsto P(k) \in \mathbb{R}^{n \times n}$ *such that*

$$A(k)P(k)A^T(k) - P(k + 1) = 0,$$

$$k \in \{k_0, k_0 + 1, \ldots, k_0 + N - 1\}, \qquad (5.8a)$$

$$P(k_0) = R^{-1}, \qquad (5.8b)$$

$$P(k) < \Gamma^{-1}(k), \quad k \in \{k_0 + 1, \ldots, k_0 + N\}. \qquad (5.8c)$$

(iv) *There exists a symmetric matrix-valued sequence* $P(\cdot) : k \in \{k_0, \ldots,$
$k_0 + N\} \mapsto P(k) \in \mathbb{R}^{n \times n}$ *such that*

$$A(k)P(k)A^T(k) - P(k + 1) < 0, \quad k \in \{k_0, \ldots, k_0 + N - 1\}, \qquad (5.9a)$$

$$P(k_0) > R^{-1}, \qquad (5.9b)$$

$$P(k) < \Gamma^{-1}(k), \quad k \in \{k_0 + 1, \ldots, k_0 + N\}. \qquad (5.9c)$$

(v) *There exists a symmetric matrix-valued sequence* $Q(\cdot) : k \in \{k_0, \ldots,$
$k_0 + N\} \mapsto Q(k) \in \mathbb{R}^{n \times n}$ *such that*

$$A^T(k)Q(k + 1)A(k) - Q(k) < 0, \quad k \in \{k_0, \ldots, k_0 + N - 1\}, \qquad (5.10a)$$

$$Q(k_0) < R, \qquad (5.10b)$$

$$Q(k) > \Gamma(k), \quad k \in \{k_0 + 1, \ldots, k_0 + N\}. \qquad (5.10c)$$

$$\square$$

Remark 5.1 The proof of the equivalence of (i) and (ii) can be found in [9, 18]. In principle, differently from the CT-LTV case, condition (ii) can be used to evaluate the FTS of system (5.2); however, it is not useful for design purposes. The fact that (iv) implies FTS is proven in [9, 18]; finally, the fact that FTS implies (iv) and the equivalence of (iii) and (iv) is proved here for the first time.

Proof $\boxed{(i) \Leftrightarrow (ii)}$ First, we prove that (ii) implies (i). Let $k \in \{k_0 + 1, \ldots, k_0 + N\}$ and $x(0)$ be such that $x^T(k_0)Rx(k_0) \leq 1$. We have

$$x(k) = \Phi(k, k_0)x(k_0).$$

Then

$$x^T(k)\Gamma(k)x(k) = x^T(k_0)\Phi^T(k, k_0)\Gamma(k)\Phi(k, k_0)x(k_0)$$

$$< x^T(k_0)Rx(k_0)$$

$$\leq 1 \quad \text{for all } k \in \{k_0 + 1, \ldots, k_0 + N\}.$$

Conversely, assume by contradiction that system (5.2) is finite-time stable and that for some $k \in \{k_0 + 1, \ldots, k_0 + N\}$ and $\bar{x} \in \mathbb{R}^n$,

$$\bar{x}^T \Phi^T(k, k_0)\Gamma(k)\Phi(k, k_0)\bar{x} \geq \bar{x}^T R\bar{x}. \tag{5.11}$$

Now let $x(k_0) = \lambda\bar{x}$, so that

$$x^T(k_0)Rx(k_0) = \lambda^2 \bar{x}^T R\bar{x} = 1; \tag{5.12}$$

moreover, let $x(\cdot)$ be the state evolution of system (5.2) starting from $x(k_0)$.
From (5.11) and (5.12) we have

$$x^T(k)\Gamma(k)x(k) = x^T(k_0)\Phi^T(k, k_0)\Gamma(k_0)\Phi(k, k_0)x(k_0)$$

$$\geq x^T(k_0)Rx(k_0) = 1.$$

Therefore, we have found an initial condition $x(k_0)$ satisfying $x^T(k_0)Rx(k_0) = 1$ such that, for some k, $x^T(k)\Gamma(k)x(k) \geq 1$. This contradicts the hypothesis that the system is finite-time stable.

$\boxed{(i) \Leftrightarrow (iii)}$ First, note that given a positive definite matrix-valued sequence $\Gamma(\cdot)$ defined over $k \in \{k_0, \ldots, k_0 + N\}$, it is always possible to find a nonsingular matrix-valued sequence $T(\cdot)$ defined over $k \in \{k_0, \ldots, k_0 + N\}$ such that $\Gamma(k) = T^T(k)T(k)$ for all $k \in \{k_0, \ldots, k_0 + N\}$. From condition (ii) we have

$$\Phi^T(k, k_0)\Gamma(k)\Phi(k, k_0) - R < 0$$

$$\Leftrightarrow \Phi^T(k, k_0)T^T(k)T(k)\Phi(k, k_0) - R < 0$$

$$\Leftrightarrow R^{-\frac{1}{2}}\Phi^T(k,k_0)T^T(k)T(k)\Phi(k,k_0)R^{-\frac{1}{2}} - I < 0$$

$$\Leftrightarrow T(k)\Phi(k,k_0)R^{-1}\Phi^T(k,k_0)T^T(k) - I < 0 \qquad (5.13)$$

for all $k \in \{k_0 + 1, \ldots, k_0 + N\}$. Now let

$$P(k) = \Phi(k,k_0)R^{-1}\Phi^T(k,k_0). \qquad (5.14)$$

Then $P(k)$ turns out to be positive definite, and Eq. (5.13) can be equivalently written as in (5.8c). Now let us recall the following properties of the transition matrix $\Phi(k,k_0)$:

$$\Phi(k_0,k_0) = I, \qquad (5.15)$$

$$\Phi(k+1,k_0) = A\Phi(k,k_0). \qquad (5.16)$$

Now Eq. (5.8b) easily follows from (5.15), whereas Eq. (5.8a) is obtained computing $P(k+1)$ from (5.14) and using (5.16).

$\boxed{\text{(iii)} \Rightarrow \text{(iv)}}$ Let us consider the following system:

$$z(k+1) = \left(A(k) + \frac{\epsilon}{2}\right)z(k), \quad z(k_0) = x_0, \quad k \in \{k_0, \ldots, k_0 + N - 1\}, \quad (5.17)$$

where ϵ is *small enough*. By continuity arguments, if system (5.2) is finite-time stable, then also system (5.17) is finite-time stable. Hence, there exists a matrix-valued sequence P_ϵ such that

$$A(k)P_\epsilon(k)A^T(k) + \epsilon P_\epsilon(k) - P_\epsilon(k+1) = 0, \quad k \in \{k_0, \ldots, k_0 + N - 1\}, \quad (5.18a)$$

$$P_\epsilon(k_0) = R^{-1}, \qquad (5.18b)$$

$$T(k)P_\epsilon(k)T^T(k) < I, \quad k \in \{k_0 + 1, \ldots, k_0 + N\}. \qquad (5.18c)$$

Exploiting again continuity arguments, there exists a scalar $\alpha > 1$ such that

$$\alpha T(k)P_\epsilon(k)T^T(k) < I, \quad k \in \{k_0 + 1, \ldots, k_0 + N\}. \qquad (5.19)$$

Letting $X(k) = \alpha P_\epsilon(k)$, Eq. (5.19) reads

$$T(k)X(k)T^T(k) < I, \quad k \in \{k_0 + 1, \ldots, k_0 + N\}, \qquad (5.20)$$

whereas (5.18a) becomes

$$A(k)X(k)A^T(k) - X(k+1) < 0, \quad k \in \{k_0, \ldots, k_0 + N - 1\},$$

since $X(k) > 0$. Finally, (5.18c) becomes

$$X(k_0) > R^{-1}.$$

$\boxed{\text{(iv)} \Leftrightarrow \text{(v)}}$ By using Schur complements, we have that (5.9a) is equivalent to

$$\begin{pmatrix} -P(k+1) & A(k)P(k) \\ P(k)A^T(k) & -P(k) \end{pmatrix} < 0. \tag{5.21}$$

By pre- and post-multiplying the last inequality by

$$\begin{pmatrix} P^{-1}(k+1) & 0 \\ 0 & P^{-1}(k) \end{pmatrix},$$

we obtain that (5.21) can be equivalently rewritten as

$$\begin{pmatrix} -P^{-1}(k+1) & P^{-1}(k+1)A(k) \\ A^T(k)P^{-1}(k+1) & -P^{-1}(k) \end{pmatrix} < 0. \tag{5.22}$$

The equivalence of (5.10a) and (5.9a) follows by letting $Q(k) = P^{-1}(k)$ and using Schur complements again, whereas one immediately sees that (5.10b) and (5.10c) are equivalent to (5.9b) and (5.9c), respectively.

$\boxed{\text{(v)} \Rightarrow \text{(i)}}$ Let us consider the quadratic function $V(x,k) = x^T(k)Q(k)x(k)$. Given a system trajectory, the Lyapunov difference $\Delta V(x,k)$ reads

$$\begin{aligned} \Delta V(x,k) &= V(x,k+1) - V(x,k) \\ &= x^T(k+1)Q(k+1)x(k+1) - x^T(k)Q(k)x(k) \\ &= x^T(k)\big(A^T(k)Q(k+1)A(k) - Q(k)\big)x(k), \end{aligned}$$

which is negative definite since (5.10a) holds. This implies that $V(x,k)$ is strictly decreasing along the trajectories of (5.1a)–(5.1b). Hence, considering an initial state x_0 such that $x_0^T R x_0 \leq 1$, we have for all $k \in \{k_0, \dots, k_0 + N\}$:

$$\begin{aligned} x(k)^T \Gamma(k)x(k) &< x(k)^T Q(k)x(k) \quad \text{by (5.10c)} \\ &< x_0^T Q(k_0)x_0 \\ &< x_0^T R x_0 \leq 1 \quad \text{by (5.10b),} \end{aligned}$$

which implies that system (5.2) is finite-time stable with respect to $(k_0, N, R, \Gamma(\cdot))$. $\Diamond$

5.4 Controller Design

Let us start with the state feedback Problem 5.1. The solution of this problem is given by the following theorem.

Theorem 5.2 (FTS of DT-LTV Systems Via State Feedback) *Problem* 5.1 *is solvable if and only if there exist a symmetric matrix-valued sequence* $P(\cdot)$ *and a matrix-valued sequence* $L(\cdot)$ *such that*

$$\begin{pmatrix} -P(k+1) & A(k)P(k) + B(k)L(k) \\ P(k)A^T(k) + L^T(k)B^T(k) & -P(k) \end{pmatrix} < 0,$$

$$k \in \{k_0, \dots, k_0 + N - 1\}, \tag{5.23a}$$

$$P(k_0) > R^{-1}, \tag{5.23b}$$

$$P(k) < \Gamma^{-1}(k), \quad k \in \{k_0 + 1, \dots, k_0 + N\}. \tag{5.23c}$$

In this case, the gain of a state feedback controller solving Problem 5.1 *is given by* $G(k) = L(k)P^{-1}(k)$. ☐

Proof First of all, note that condition (5.8a) can be equivalently rewritten as

$$\begin{pmatrix} -P(k+1) & A(k)P(k) \\ P(k)A^T(k) & -P(k) \end{pmatrix} < 0, \quad k \in \{k_0, \dots, k_0 + N - 1\}. \tag{5.24}$$

Now we can apply the last inequality to system (5.6), by replacing $A(k)$ with $A(k) + B(k)G(k)$; in this way, we find that the system is guaranteed to be finite-time stable wrt $(k_0, N, R, \Gamma(\cdot))$ if and only if

$$\begin{pmatrix} -P(k+1) & (A(k) + B(k)G(k))P(k) \\ P(k)(A(k) + B(k)G(k))^T & -P(k) \end{pmatrix} < 0,$$

$$k \in \{k_0, \dots, k_0 + N - 1\},$$

$$P(k_0) > R^{-1},$$

$$P(k) < \Gamma^{-1}(k), \quad k \in \{k_0 + 1, \dots, k_0 + N\}.$$

The proof follows by letting $G(k)P(k) = L(k)$. ◊

The numerical implementation of Theorem 5.2 will be discussed in Sect. 5.5.

Next, we move to the finite-time stabilizability via output feedback. The sufficiency of the conditions proven in the following theorem can be found in [9, 18]; the necessity is proven here for the first time.

Theorem 5.3 (FTS of DT-LTV Systems Via Output Feedback) *Problem* 5.2 *is solvable if and only if there exist symmetric matrix-valued sequences* $Q(\cdot)$ *and* $S(\cdot)$, *a nonsingular matrix-valued sequence* $U(\cdot)$, *and matrix-valued sequences* $\hat{A}_K(\cdot)$, $\hat{B}_K(\cdot)$, $\hat{C}_K(\cdot)$, *and* $D_K(\cdot)$ *such that*

$$\begin{pmatrix} \Theta_a(k) & \Theta_b(k) \\ \Theta_b^T(k) & \Theta_a(k+1) \end{pmatrix} < 0, \quad k \in \{k_0, \dots, k_0 + N - 1\}, \tag{5.25a}$$

$$\begin{pmatrix} Q(k) & \Psi_{12}(k) & \Psi_{13}(k) & \Psi_{14}(k) \\ \Psi_{12}^T(k) & \Psi_{22}(k) & 0 & 0 \\ \Psi_{13}^T(k) & 0 & I & 0 \\ \Psi_{14}^T(k) & 0 & 0 & I \end{pmatrix} > 0, \quad k \in \{k_0 + 1, \dots, k_0 + N\},$$

(5.25b)

$$\begin{pmatrix} Q(k_0) & I \\ I & S(k_0) \end{pmatrix} < \begin{pmatrix} \Delta_a & Q(k_0)R \\ RQ(k_0) & R \end{pmatrix},$$

(5.25c)

where

$$\Theta_a(k) = -\begin{pmatrix} Q(k) & I \\ I & S(k) \end{pmatrix},$$

(5.26a)

$$\Theta_b(k) = \begin{pmatrix} Q(k)A^T(k) + \hat{C}_K^T(k)B^T(k) & \hat{A}_K^T(k) \\ A^T(k) + C^T(k)D_K^T(k)B^T(k) & A^T(k)S(k+1) + C^T(k)\hat{B}_K^T(k) \end{pmatrix},$$

(5.26b)

$$\Psi_{12}(k) = I - Q(k)\Gamma(k),$$

(5.26c)

$$\Psi_{13}(k) = Q(k)\Gamma^{1/2}(k),$$

(5.26d)

$$\Psi_{14}(k) = U(k)\Gamma_K^{1/2}(k),$$

(5.26e)

$$\Psi_{22}(k) = S(k) - \Gamma(k),$$

(5.26f)

$$\Delta_a = Q(0)RQ(0) + U(0)R_K U^T(0).$$

(5.26g)

$\square$

Proof The connection between system (5.1a)–(5.1b) and controller (5.7a)–(5.7b) reads

$$\begin{pmatrix} x(k+1) \\ x_c(k+1) \end{pmatrix} = A_{\mathrm{CL}}(k) \begin{pmatrix} x(k) \\ x_c(k) \end{pmatrix},$$

(5.27)

where

$$A_{\mathrm{CL}}(k) := \begin{pmatrix} A(k) + B(k)D_K(k)C(k) & B(k)C_K(k) \\ B_K(k)C(k) & A_K(k) \end{pmatrix}.$$

Define $R_{\mathrm{CL}} = \mathrm{diag}(R, R_K)$ and $\Gamma_{\mathrm{CL}}(k) = \mathrm{diag}(\Gamma(k), \Gamma_K(k))$. From (v) in Theorem 5.1 it follows that Problem 5.2 is solvable if and only if there exist a positive definite matrix-valued sequence $X(\cdot)$ and matrices $A_K(\cdot)$, $B_K(\cdot)$, $C_K(\cdot)$, and $D_K(\cdot)$

such that

$$\begin{pmatrix} -X(k) & A_{\mathrm{CL}}^T(k)X(k+1) \\ X(k+1)A_{\mathrm{CL}}(k) & -X(k+1) \end{pmatrix} < 0, \quad k \in \{k_0, \dots, k_0 + N - 1\},$$

$$\text{(5.28a)}$$

$$X(k) > \Gamma_{\mathrm{CL}}(k), \quad k \in \{k_0 + 1, \dots, k_0 + N\}, \tag{5.28b}$$

$$X(0) < R_{\mathrm{CL}}(k). \tag{5.28c}$$

Now, according to Lemma 3.1, let us define

$$X(k) = \begin{pmatrix} S(k) & M(k) \\ M^T(k) & \star \end{pmatrix}, \qquad X^{-1}(k) = \begin{pmatrix} Q(k) & U(k) \\ U^T(k) & \star \end{pmatrix},$$

where $\star$ denotes a "do not care" block, and

$$\Pi_1(k) = \begin{pmatrix} Q(k) & I \\ U^T(k) & 0 \end{pmatrix}, \qquad \Pi_2(k) = \begin{pmatrix} I & S(k) \\ 0 & M^T(k) \end{pmatrix}.$$

Note that, by definition,

$$S(k)Q(k) + M(k)U^T(k) = I, \tag{5.29a}$$

$$Q(k)S(k) + U(k)M^T(k) = I, \tag{5.29b}$$

$$X(k)\Pi_1(k) = \Pi_2(k). \tag{5.29c}$$

By pre and post-multiplying inequality (5.28a) by $\mathrm{diag}(\Pi_1^T(k), \Pi_1^T(k+1))$ and $\mathrm{diag}(\Pi_1(k), \Pi_1(k+1))$, respectively, pre- and post-multiplying (5.28b) and (5.28c) by $\Pi_1^T(k)$ and $\Pi_1(k)$, respectively, and taking into account (5.29a)–(5.29c) and Lemma 3.1, the proof follows once we let

$$\hat{B}_K(k) = M(k+1)B_K(k) + S(k+1)B(k)D_K(k), \tag{5.30a}$$

$$\hat{C}_K(k) = C_K(k)U^T(k) + D_K(k)C(k)Q(k), \tag{5.30b}$$

$$\begin{aligned} \hat{A}_K(k) = {} & M(k+1)A_K(k)U^T(k) + S(k+1)B(k)C_K(k)U^T(k) \\ & + M(k+1)B_K(k)C(k)Q(k) \\ & + S(k+1)\big(A(k) + B(k)D_K(k)C(k)\big)Q(k). \end{aligned} \tag{5.30c}$$

Note that (5.25a) implies that, at each time instant k,

$$\begin{pmatrix} Q(k) & I \\ I & S(k) \end{pmatrix} > 0, \tag{5.31}$$

which, according to Lemma 3.1, guarantees the reconstruction of $X(k)$ starting from the knowledge of $S(k)$, $Q(k)$, and $U(k)$. $\quad\diamond$

Remark 5.2 (Controller Design)   Assume now that the hypotheses of Theorem 5.3 are satisfied. In order to design the controller, the following steps have to be followed:

(i) Find $Q(\cdot)$, $S(\cdot)$, $U(\cdot)$, $\hat{A}_K(\cdot)$, $\hat{B}_K(\cdot)$, $\hat{C}_K(\cdot)$, and $D_K(\cdot)$ such that (5.25a)–(5.25c) are satisfied.
(ii) Find a matrix-valued sequence $M(\cdot)$ such that $M(k) = (I - S(k)Q(k))U^{-T}(k)$.
(iii) Obtain $A_K(\cdot)$, $B_K(\cdot)$, $C_K(\cdot)$, and $D_K(\cdot)$ by inverting (5.30a)–(5.30c). The controller matrices read

$$B_K(k) = M^{-1}(k+1)\hat{B}_K(k) - M^{-1}(k+1)S(k+1)B(k)D_K(k),$$

$$C_K(k) = \hat{C}_K(k)U^{-T}(k) - D_K(k)C(k)Q(k)U^{-T}(k),$$

$$A_K(k) = M^{-1}(k+1)\hat{A}_K(k)U^{-T}(k) - M^{-1}(k+1)S(k+1)B(k)C_K(k)$$
$$- B_K(k)C(k)Q(k)U^{-T}(k) - M^{-1}(k+1)S(k+1)$$
$$\times \big(A(k) + B(k)D_K(k)C(k)\big)Q(k)U^{-T}(k).$$

The procedure for the design of the output feedback controller expressed in terms of solution of a set of LMIs will be presented in next section.

5.5 Numerical Implementation of the Main Results

In this section, we will discuss the numerical implementation of the two main results of the paper, namely Theorem 5.2, regarding the state feedback design, and Theorem 5.3, regarding the output feedback design. We will show that in both cases, it is possible to express the conditions in terms of a feasibility problem involving a set of LMIs; in this way, the solution, if it exists, can be found using one of the many packages that are available to solve such problems. However, in the case of the output feedback, the proposed implementation introduces some conservativeness, and therefore the conditions are no longer necessary but only sufficient.

Conditions (5.23a)–(5.23c) of Theorem 5.2 are already expressed in terms of LMIs with $L(\cdot)$ and $Q(\cdot)$ as optimization variables. Therefore, once $(k_0, N, R, \Gamma(\cdot))$ have been fixed, the design of the state feedback controller is expressed as an LMI feasibility problem with a set of $(2N + 1)$ LMIs with $(2N + 1)$ optimization variables $(L(k_0), \ldots, L(N - 1), Q(k_0), \ldots, Q(N))$. This feasibility problem can be efficiently tackled with the Matlab Robust Control Toolbox [75].

Now let us move to the output feedback design. Conditions (5.25a) and (5.25b) are already expressed as DLMI. Conversely, the initial condition (5.25c) exhibits a quadratic term in the optimization variables $Q(k_0)$ and $U(k_0)$. In order to replace condition (5.25c) by an LMI, a procedure analogous to that developed in Sect. 3.3 can be used.

Eventually, in order to compute $M(\cdot)$, the matrix-valued sequence $U(\cdot)$ has to be invertible. To this end, we can enforce the condition $U(k) > 0$ for all $k \in \{k_0, k_0 + 1, \ldots, k_0 + N\}$ since Lemma 3.2 also holds in the discrete-time case.

Fig. 5.1 The mechanical system considered in Example 5.1

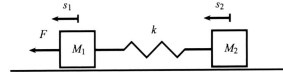

Therefore, also in the case of the output feedback design, we end up with a set of LMIs. More specifically, once $(k_0, N, \mathrm{diag}(R, R_K), \mathrm{diag}(\Gamma(\cdot), \Gamma_K(\cdot)))$ have been fixed, we need to check the feasibility of a set of $(3N + 2)$ LMIs expressed in terms of $(7N + 3)$ optimization variables: the $4N$ matrices $\hat{A}(k)$, $\hat{B}(k)$, $\hat{C}(k)$, $D(k)$ and the $3(N + 1)$ matrices $Q(k_0), \dots, Q(k_0 + N)$, $S(k_0), \dots, S(k_0 + N)$, $U(k_0), \dots, U(k_0 + N)$.

Example 5.1 Let us consider the mechanical system shown in Fig. 5.1 and consisting of two masses linked by a spring. The masses slide on a plane without any friction. Letting $x = (s_1\ \dot{s}_1\ s_2\ \dot{s}_2)^T$, $u = F$, $y_1 = s_1$, $y_2 = s_2$ and supposing $M_1 = M_2 = 1$ kg and $k = 1$ N/m, the equations of the system are

$$\dot{x} = A_c x + B_c u, \tag{5.32a}$$

$$y = Cx, \tag{5.32b}$$

where

$$A_c = \begin{pmatrix} 0 & 1 & 0 & 0 \\ -1 & 0 & 1 & 0 \\ 0 & 0 & 0 & 1 \\ 1 & 0 & -1 & 0 \end{pmatrix}, \qquad B_c = \begin{pmatrix} 0 \\ 1 \\ 0 \\ 0 \end{pmatrix}, \qquad C = \begin{pmatrix} 1 & 0 & 0 & 0 \\ 0 & 0 & 1 & 0 \end{pmatrix}.$$

The discrete-time version of system (5.32a)–(5.32b) obtained using the ZOH method is given by

$$x(k + 1) = A_d x(k) + B_d u(k), \tag{5.33a}$$

$$y(k) = Cx(k), \tag{5.33b}$$

where

$$A_d = \exp(A_c T), \qquad B_d = \int_0^T \exp(A_c \sigma)\, d\sigma\, B_c.$$

The sampling time was chosen $T = 0.2$ s, which is about 1/20th of the period of system's oscillating modes. In this case, the A_d and B_d matrices are approximately given by

$$A_d = \begin{pmatrix} 0.98 & 0.20 & 0.02 & 0.00 \\ -0.20 & 0.98 & 0.20 & 0.02 \\ 0.02 & 0.00 & 0.98 & 0.20 \\ 0.20 & 0.02 & -0.20 & 0.98 \end{pmatrix}, \qquad B_d = \begin{pmatrix} 0.02 \\ 0.20 \\ 0.00 \\ 0.00 \end{pmatrix}.$$

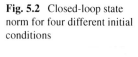

Fig. 5.2 Closed-loop state norm for four different initial conditions

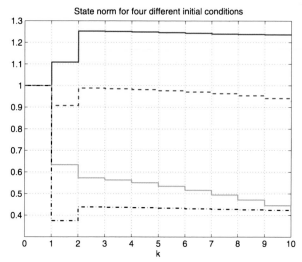

Our goal is to solve Problem 5.2 with $k_0 = 0$, $N = 10$, $R = R_K = I$, and $\Gamma(k) = \Gamma_K(k) = I/(1.3)^2$. The system under investigation is not asymptotically stable because of the absence of any friction; the A_d matrix exhibits all its eigenvalues placed on the unit circle.

Using the Robust Control Toolbox [75], it is possible to prove that this problem is feasible and an output feedback controller can be designed. In Fig. 5.2 four sample evolutions from different initial conditions are presented, showing that the designed controller guarantees the FTS of the closed-loop system. These initial conditions have been obtained from a random sampling over 500000 samples.

5.6 Summary

In this chapter, we have dealt with the finite-time control of DT-LTV systems. First, some necessary and sufficient conditions for FTS have been provided in terms of either the state transition matrix, or the existence of a solution to a certain LDE, or the existence of a feasible solution to a coupled DLMI/LMIs. Starting from the DLMI-based condition, necessary and sufficient condition for the existence of state and output feedback controllers guaranteeing the finite-time stabilization of the closed-loop system have been derived.

The condition for the state feedback design can be immediately translated into an LMI-based feasibility problem, while the output feedback condition cannot, since it is necessary to replace, at the price of some conservativeness, the initial condition, involving the optimization variables, into an LMI constraint.

In the following, we shall briefly outline a different approach, specifically developed for the finite-time control of discrete-time linear time-invariant (DT-LTI) systems; such an approach is based on time-invariant Lyapunov functions, rather than on time-varying Lyapunov functions. Obviously, the obtained results turn out to be more conservative than those stated in Sects. 5.3–5.4 since they are only sufficient conditions for FTS. The following of the section is adapted from [2].

First, let us restrict Definition 5.1 to the case of DT-LTI systems, assuming that both the initial set and the trajectory set are time-invariant ellipsoids having the same shape. For DT-LTI systems, the FTS definition (5.1) particularizes as follows.

Definition 5.2 (FTS of DT-LTI Systems with Ellipsoidal Domains) Given an integer N and positive definite matrices R and Γ with $\Gamma > R$, the DT-LTI system

$$x(k+1) = Ax(k), \quad x(0) = x_0, \tag{5.34}$$

is said to be finite-time stable with respect to (T, R, Γ) if

$$x_0^T R x_0 \le 1 \Rightarrow x(k)^T \Gamma x(k) < 1, \quad k \in \{1, \ldots, N\}. \tag{5.35}$$

$$\Diamond$$

The following theorem is derived under the assumption that the initial set and the trajectory set have the same shape, namely $\Gamma = \rho R$, for a given positive scalar $\rho < 1$.

Theorem 5.4 [2] *System* (5.34) *is finite-time stable with respect to* $(T, R, \rho R)$ *with* $0 < \rho < 1$ *if, letting* $\tilde{P} = R^{-\frac{1}{2}} P R^{-\frac{1}{2}}$, *there exist a scalar* $\gamma \ge 1$ *and a positive definite matrix* $P \in \mathbb{R}^{n \times n}$ *such that*

$$A^T P A - \gamma P < 0, \tag{5.36a}$$

$$\mathrm{cond}(\tilde{P}) < \frac{1}{\rho} \gamma^{-N}. \tag{5.36b}$$

$$\Box$$

To prove Theorem 5.4, let us assume that $x^T(0) R x(0) \le 1$. We want to prove that if conditions (5.36a)–(5.36b) hold, then $x^T(k) \Gamma x(k) < 1$ for all $k = 1, \ldots, N$. Let $V(x) = x^T P x$; simple calculations show that (5.36a) implies

$$V(x(k+1)) < \gamma V(x(k)), \quad k \in \{0, \ldots, N\}, \tag{5.37}$$

where $x(k+1)$ is evaluated along the solutions of (5.34). Applying iteratively (5.37) we obtain

$$V(x(k)) < \gamma^k V(x(0)), \quad k \in \{1, \ldots, N\}. \tag{5.38}$$

Now, letting $\tilde{P} = R^{-1/2} P R^{-1/2}$ and using the fact that $\gamma \ge 1$, we have

$$\gamma^k V(x(0)) \le \gamma^N \lambda_{\max}(\tilde{P}) \tag{5.39}$$

and

$$V(x(k)) \ge \lambda_{\min}(\tilde{P}) x^T(k) R x(k). \tag{5.40}$$

Putting together (5.38)–(5.40), we obtain

$$x^T(k)Rx(k) < \frac{\lambda_{\max}(\tilde{P})}{\lambda_{\min}(\tilde{P})} \gamma^N. \qquad (5.41)$$

From (5.41) it follows that (5.36b) implies that

$$x^T(k)\rho Rx(k) < 1, \quad k = 1, \dots, N.$$

From Theorem 5.4 it readily follows a condition for the existence of a state feedback controller such that the closed-loop system is finite-time stable.

Corollary 5.1 (Finite-Time Stabilization of DT-LTI Systems Via State Feedback) *System (5.34) is finite-time stabilizable via a state feedback control law of the form $u(k) = Kx(k)$ with respect to $(T, R, \rho R)$ if there exist a positive definite matrix $Q \in \mathbb{R}^{n \times n}$, a matrix $L \in \mathbb{R}^{m \times n}$, and a scalar $\gamma \geq 1$ such that*

$$\begin{pmatrix} -\gamma Q & (AQ + BL)^T \\ AQ + BL & -Q \end{pmatrix} < 0, \qquad (5.42a)$$

$$\mathrm{cond}(\tilde{Q}) < \frac{1}{\rho}\gamma^{-N}, \qquad (5.42b)$$

where $\tilde{Q} = R^{1/2}QR^{1/2}$. In this case, the controller K is given by $K = LQ^{-1}$.

Finally, by following the same guidelines of Sect. 3.5, a condition for finite-time stabilization via output feedback of DT-LTI systems, consistent with Definition 5.2, can be derived.

Chapter 6
FTS Analysis Via PQLFs

6.1 Introduction

In the previous chapters, the definition of FTS made use of the standard weighted quadratic norm to define both the initial domain and the trajectory domain. Therefore, such domains turned out to be ellipsoidal. The definition of the above domains is consistent with the fact that quadratic Lyapunov functions were used to derive the main results for both the FTS analysis and the finite time stabilization problem. In [49] some necessary and sufficient conditions for FTS have been proposed for the case of polytopic trajectory domain, while the initial domain is still assumed to be ellipsoidal.

The use of polytopes, rather that ellipsoids, is important to tackle many practical problems where, for instance, the constraints on the state variables are of the form $a_i \leq x_i \leq b_i$, $i = 1, \dots, n$. Although polytopes can be always approximated by ellipsoids, the FTS conditions for ellipsoidal domains turn out to be conservative in this case (see [16]). In [16] ad hoc conditions for the FTS with polytopic domains, based on the use of polyhedral Lyapunov functions, have been proposed. However, these conditions were implemented via a procedure that, due to the nonconvexity of the optimization function, required a high computational burden.

In this chapter, the class of PQLFs for the FTS analysis of CT-LTV systems is considered. This class of functions has been widely adopted for the stability analysis and control design of piecewise linear systems [60, 79], i.e., dynamical systems that exhibit different linear dynamics depending on the region the state variables belong to.

To fit the PQLF-based machinery in the FTS context, the concept of *PQLF defined over a conical partition* is introduced. This class of functions can be obtained by partitioning the state space into conic regions; then a quadratic form is associated to each cone. While in [60, 79] the number and the shapes of the partitions were only related to the system dynamics, in our approach, it is a degree of freedom used to define the level curves of the Lyapunov function.

The proposed framework allows us to find computationally efficient conditions for the FTS of linear systems when the initial and trajectory domains are piecewise

F. Amato et al., *Finite-Time Stability and Control*,
Lecture Notes in Control and Information Sciences 453,
DOI 10.1007/978-1-4471-5664-2_6, © Springer-Verlag London 2014

quadratic, i.e., their boundaries are the level curves of PQLFs. Obviously, this class of domains includes the class of ellipsoids, and, at the same time, it also represents a meaningful generalization of polytopic domains. Numerical examples confirm that this approach turns out to be less conservative than the results available when poly-topic domains are considered. Moreover, the technique is much more efficient from the numerical point of view since the main result requires the solution of a feasibility problem involving DLMIs, which can be solved via off-the-shelf convex optimization algorithms. Finally, our approach allows us to tackle the more general case where the initial and/or trajectory domains are piecewise quadratic.

6.2 Preliminaries

6.2.1 Cones and Conical Partitions

In this subsection, we introduce some auxiliary concepts in order to define the class of PQLFs and the corresponding class of Piecewise Quadratic Domains (PQDs) over conical partitions.

In the following, we say that a set S is a (convex) cone in $\mathbb{R}^n$ if it is generated through the *conic* combination of p vectors, $x_i \in \mathbb{R}^n$, $i = 1, \ldots, p$, i.e.,

$$S = \mathrm{cone}(\{x_1, \ldots, x_p\}) := \left\{ x : x = \sum_{i=1}^{p} \lambda_i x_i, \lambda_i \geq 0, i = 1, \ldots, p \right\}. \qquad (6.1)$$

Let S be a cone in $\mathbb{R}^n$, defined according to (6.1); the *dimension* of S is the column rank of the matrix $\mathcal{M}_S := [x_1 \ \ldots \ x_p]$, that is,

$$\dim(S) = \mathrm{rank}(\mathcal{M}_S) \leq n.$$

Given a cone S, defined according to (6.1), any minimal set of $q \leq p$ vectors $\{\hat{x}_1, \ldots, \hat{x}_q\}$, with $\|\hat{x}_i\| = 1, i = 1, \ldots, q$, such that

$$S = \mathrm{cone}(\{\hat{x}_1, \ldots, \hat{x}_q\}), \qquad (6.2)$$

is said to be a set of *normalized extremal rays* of S.

For the purpose of the theory developed in this chapter, we give the following definition.

Definition 6.1 (Regular Cones) A cone S is said to be *regular* if the set of normalized extremal rays is univocally determined; in this case, such a set is denoted by $\mathrm{extr}(S)$.   ◇

Remark 6.1 Definition 6.1 is needed since there are particular cases that are not of interest for the theory developed in this chapter when a cone admits infinitely many sets of normalized extremal rays. One example is represented by the whole space $\mathbb{R}^2$, which is a cone of dimension 2 and is generated by any set of normalized extremal rays of the form $\{x_1, x_2, x_3\}$ where $x_1 = (\sqrt{2}/2 \; \sqrt{2}/2)^T$, $x_2 = (-\sqrt{2}/2 \; \sqrt{2}/2)^T$, and $x_3 = (h \; k)^T$ with $k < 0$, $|h| < |k|$, and $\|x_3\| = 1$. It is worth noting that nonregular cones are not useful to forming conical partitions, which will be considered later. Such conical partitions will be composed only by regular cones; this in turn guarantees that the set $\text{extr}(S)$ is well defined for each cone forming the partition. $\diamond$

Now we are ready to introduce the concept of conical partition of the space $\mathbb{R}^n$.

Definition 6.2 (Conical Partition of $\mathbb{R}^n$) A conical partition $\mathcal{P} = \{S_1, \ldots, S_r\}$ of $\mathbb{R}^n$ is a collection of cones S_i, $i = 1, \ldots, r$, such that:

- each cone S_i, $i = 1, \ldots, r$, is regular;
- each cone S_i has dimension n, i.e., $\dim\{S_i\} = n$, $i = 1, \ldots, r$;
- the union of the cones S_i, $i = 1, \ldots, r$, covers the whole state space $\mathbb{R}^n$, that is, $\bigcup_{i=1}^{r} S_i = \mathbb{R}^n$;
- for $i \neq j$, the intersection of the interiors of S_i and S_j is empty:

$$\text{int}\{S_i\} \cap \text{int}\{S_j\} = \emptyset. \tag{6.3}$$

$\diamond$

The *set of generating rays*, say $\mathcal{R}_\mathcal{P}$, of a conical partition $\mathcal{P}$ is defined as the union of the sets of the normalized extremal rays of each cone in $\mathcal{P}$, i.e., $\mathcal{R}_\mathcal{P} = \bigcup_{S_i \in \mathcal{P}} \text{extr}(S_i) = \{\hat{x}_1, \ldots, \hat{x}_v\}$. Note that the cardinality of $\mathcal{R}_\mathcal{P}$, namely v, depends on both the dimension of the state space, n, and the number of cones, r.

In the following, we will denote the *intersection of two cones* as

$$H(S_i, S_j) := S_i \cap S_j.$$

For two distinct cones S_i, S_j in a same partition $\mathcal{P}$, from (6.3) it follows that

$$H(S_i, S_j) = \text{cone}\big\{\text{extr}\{S_i\}\big\} \cap \text{cone}\big\{\text{extr}\{S_j\}\big\}$$

$$= \text{cone}\big\{\text{extr}\{S_i\} \cap \text{extr}\{S_j\}\big\}, \tag{6.4}$$

where $\dim(H(S_i, S_j)) \leq n - 1$.

Definition 6.3 (Union of Conical Partitions of $\mathbb{R}^n$) Given two conical partitions $\mathcal{P}^1$ and $\mathcal{P}^2$ of $\mathbb{R}^n$, let us define the union of $\mathcal{P}^1$ and $\mathcal{P}^2$ as the conical partition $\mathcal{P}$ whose set of generating rays is given by

$$\mathcal{R}_\mathcal{P} = \mathcal{R}_{\mathcal{P}^1} \cup \mathcal{R}_{\mathcal{P}^2}. \tag{6.5}$$

$\diamond$

An example of union of partitions in the two-dimensional case is illustrated in Fig. 6.1.

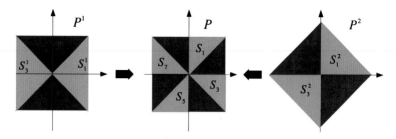

Fig. 6.1 The partition $\mathcal{P} = \{S_1, \ldots, S_8\}$ of $\mathbb{R}^2$ is defined as the union of $\mathcal{P}^1 = \{S_1^1, S_2^1, S_3^1, S_4^1\}$ and $\mathcal{P}^2 = \{S_1^2, S_2^2, S_3^2, S_4^2\}$

6.2.2 PQLFs and PQDs Defined over a Conical Partition

In this section, we introduce the class of PQLFs and the corresponding PQDs defined over a partition $\mathcal{P}$.

Definition 6.4 (PQLFs over a Conical Partition) A time-varying PQLF, defined over a conical partition $\mathcal{P} = \{S_1, \ldots, S_r\}$ of $\mathbb{R}^n$, is a space-continuous, piecewise continuously time-differentiable, and positive definite function of the form

$$F_{\mathcal{P}}(x, t) = x^T F_i(t)x, \quad x \in S_i, \quad i = 1, \ldots, r, \tag{6.6}$$

where $F_i(t) \in \mathbb{R}^{n \times n}$, $i = 1, \ldots, r$, are symmetric matrix-valued functions positive definite in the cone S_i, that is, for $t \geq 0$,

$$x^T F_i(t)x > 0, \quad x \in S_i - \{0\}, \quad i = 1, \ldots, r. \tag{6.7}$$

$\Diamond$

Note that, given a PQLF $F_{\mathcal{P}}(\cdot, \cdot)$, the subscript takes into account the partition on which the PQLF is defined, while F recalls the set of matrix functions, $F_i(\cdot)$, $i = 1, \ldots, r$, attained by the PQLF itself over the partition.

In order to ensure the space-continuity of the PQLF $F_{\mathcal{P}}(x, t)$, the condition

$$x^T F_i(t)x = x^T F_j(t)x, \quad x \in H(S_i, S_j), \tag{6.8}$$

has to be satisfied for $i, j = 1, \ldots, r, i \neq j, t \geq 0$.

According to the previous definition, it is possible to introduce the concept of PQD over a conical partition.

Definition 6.5 (PQD over a Conical Partition) A time-varying PQD, defined over a conical partition $\mathcal{P} = \{S_1, \ldots, S_r\}$ of $\mathbb{R}^n$, is a compact domain whose boundary is the unitary level curve of a PQLF $F_\mathcal{P}(x, t)$, that is,

$$\mathcal{X}_{F_\mathcal{P}}(t) := \{x : F_\mathcal{P}(x, t) \le 1\}$$
$$= \{x : x^T F_i(t)x \le 1, x \in S_i, i = 1, 2, \ldots r\}. \tag{6.9}$$

$\Diamond$

In the following, we shall denote a time-invariant PQLF and a time-invariant PQD by $F_\mathcal{P}(x)$ and $\mathcal{X}_{F_\mathcal{P}}$, respectively.

Remark 6.2 Let us consider a PQD $\mathcal{X}(\cdot)$ defined via a conical partition $\mathcal{P}^1$ of $\mathbb{R}^n$ by means of a PQLF $F_{\mathcal{P}^1}(x, t)$, that is $\mathcal{X}(\cdot) = \mathcal{X}_{F_{\mathcal{P}^1}}(\cdot)$. Given a conical partition $\mathcal{P}^2$ whose set of generating rays $\mathcal{R}_{\mathcal{P}^2} \supseteq \mathcal{R}_{\mathcal{P}^1}$, the domain $\mathcal{X}(\cdot)$ can be equivalently defined via the conical partition $\mathcal{P}^2$ by means of the natural extension of the PQLF $F_{\mathcal{P}^1}(x, t)$ to the partition $\mathcal{R}_{\mathcal{P}^2}$, namely the PQLF $\tilde{F}_{\mathcal{P}^2}(x, t)$ defined in such a way that, for all x, t, we have $\tilde{F}_{\mathcal{P}^2}(x, t) = F_{\mathcal{P}^1}(x, t)$. $\Diamond$

The set of PQDs defined in (6.9) represents a generalization of the set of ellipsoidal domains; indeed, an ellipsoidal domain can be expressed as a PQD with $F_i(t) = F(t) > 0, i = 1, \ldots, r, t \ge 0$. Moreover, it includes the set of polytopic domains whose boundary is the unitary level curve of a polyhedral Lyapunov function [39], as proved in the following theorem. For the sake of simplicity, the theorem is stated for the time-invariant case. The proof exploits the concept of polyhedral Lyapunov functions of the second order proposed in [58].

Theorem 6.1 [29] *Let us consider a polyhedral Lyapunov function $F_{\text{pol}}(x)$ defined over a conical partition $\mathcal{P} = \{S_1, \ldots, S_r\}$ such that*

$$F_{\text{pol}}(x) = c_i^T x, \quad x \in S_i, \quad i = 1, \ldots, r, \tag{6.10}$$

with $c_i \in \mathbb{R}^n$, and the corresponding polytope

$$\mathcal{X}_{F_{\text{pol}}} := \{x : F_{\text{pol}}(x) \le 1\}.$$

Then, there always exists a PQLF $F_\mathcal{P}(x)$ defined over the same partition $\mathcal{P}$ such that

$$\mathcal{X}_{F_\mathcal{P}} = \mathcal{X}_{F_{\text{pol}}}. \tag{6.11}$$

$\square$

Proof The proof is constructive. Let us consider a generic cone $S_i \in \mathcal{P}$. From Eq. (6.10) it follows that, for $x \in S_i$, $F_{\text{pol}}(x) = c_i^T x$. By squaring $F_{\text{pol}}(x)$ we obtain

$$F_{\text{pol}}^2(x) = \left(c_i^T x\right)^2 = x^T \left(c_i c_i^T\right)x, \quad x \in S_i, \quad i = 1, \ldots, r.$$

It is straightforward to verify that the functions $F_{\mathrm{pol}}(x)$ and $F_{\mathrm{pol}}^2(x)$ have the same unitary level curve, and hence

$$\mathcal{X}_{F_{\mathrm{pol}}} = \mathcal{X}_{F_{\mathrm{pol}}^2}. \tag{6.12}$$

Exploiting Eq. (6.12), it is possible to build a PQLF $F_{\mathcal{P}}(x)$ satisfying (6.11) by choosing the following set of diadic matrices:

$$F_i = c_i c_i^T, i = 1, \ldots, r. \qquad\qquad \Diamond$$

To conclude this section, we restrict Definition 1.1 to the case of autonomous CT-LTV systems (2.1) assuming that both the initial and trajectory domains are PQDs.

Definition 6.6 (FTS of CT-LTV Systems with PQDs) Given

- an initial time t_0 and a positive scalar T,
- a time–invariant PQD $\mathcal{X}_{R_{\mathcal{P}^1}}$, $R_{\mathcal{P}^1}(x)$ being a time-invariant PQLF defined over the partition $\mathcal{P}^1$ of $\mathbb{R}^n$,
- a time-varying PQD $\mathcal{X}_{\Gamma_{\mathcal{P}^2}}(t)$, $\Gamma_{\mathcal{P}^2}(x,t)$ being a time-varying PQLF defined over the partition $\mathcal{P}^2$ of $\mathbb{R}^n$, with $t \in [t_0, t_0 + T]$,

system (2.1) is said to be FTS with respect to $(t_0, T, \mathcal{X}_{R_{\mathcal{P}^1}}, \mathcal{X}_{\Gamma_{\mathcal{P}^2}}(\cdot))$ if

$$x_0 \in \mathcal{X}_{R_{\mathcal{P}^1}} \Rightarrow x(t, x_0) \in \mathcal{X}_{\Gamma_{\mathcal{P}^2}}(t), \quad t \in [t_0, t_0 + T]. \tag{6.13}$$
$$\Diamond$$

6.3 FTS with PQDs

In this section, we first propose a sufficient condition for the FTS of the CT-LTV system (2.1). Such result will be further investigated to introduce some conditions that can be checked numerically in a more efficient way by means of DLMIs.

Theorem 6.2 [29] *Let us consider two PQDs $\mathcal{X}_{R_{\mathcal{P}^1}}$ and $\mathcal{X}_{\Gamma_{\mathcal{P}^2}}(t)$, $t \in [t_0, t_0 + T]$, defined over the partitions $\mathcal{P}^1 = \{S_1^1, S_2^1, \ldots, S_{r1}^1\}$ and $\mathcal{P}^2 = \{S_1^2, S_2^2, \ldots, S_{r2}^2\}$ of $\mathbb{R}^n$, respectively. Then system (2.1) is finite-time stable with respect to $(t_0, T, \mathcal{X}_{R_{\mathcal{P}^1}}, \mathcal{X}_{\Gamma_{\mathcal{P}^2}}(\cdot))$ if there exists a PQLF $P_{\mathcal{P}}(x,t)$, defined over the partition $\mathcal{P} = \{S_1, \ldots, S_r\}$, the union of $\mathcal{P}^1$ and $\mathcal{P}^2$, verifying the equality conditions (6.8) for space-continuity, namely*

$$x^T P_i(t)x = x^T P_j(t)x, \quad x \in H(S_i, S_j), \tag{6.14}$$

for $i, j = 1, \ldots, r, i \neq j, t \in [t_0, t_0 + T]$, and such that

$$x^T \left(\dot{P}_i(t) + A(t)^T P_i(t) + P_i(t) A(t) \right)x < 0, \tag{6.15a}$$

$$x^T \left(P_i(t) - \tilde{\Gamma}_i(t) \right) x \geq 0, \qquad\qquad (6.15b)$$

$$x^T \left(P_i(t_0) - \tilde{R}_i \right) x \leq 0, \qquad\qquad (6.15c)$$

for $x \in S_i$, $i = 1, \ldots, r$, $t \in [t_0, t_0 + T]$, where $\tilde{R}_{\mathcal{P}}$ and $\tilde{\Gamma}_{\mathcal{P}}(\cdot)$ are defined in such a way that

$$\mathcal{X}_{\tilde{R}_{\mathcal{P}}} = \mathcal{X}_{R_{\mathcal{P}^1}}, \qquad \mathcal{X}_{\tilde{\Gamma}_{\mathcal{P}}}(\cdot) = \mathcal{X}_{\Gamma_{\mathcal{P}^2}}(\cdot). \qquad\qquad (6.16)$$

□

Proof The partition $\mathcal{P}$ is defined as the union of $\mathcal{P}^1$ and $\mathcal{P}^2$, and hence, due to (6.5), the set $\mathcal{R}_{\mathcal{P}}$ of the generating rays of $\mathcal{P}$ contains both $\mathcal{R}_{\mathcal{P}^1}$ and $\mathcal{R}_{\mathcal{P}^2}$. According to Remark 6.2, it is possible to define the initial domain $\mathcal{X}_{R_{\mathcal{P}^1}}$ and trajectory domain $\mathcal{X}_{\Gamma_{\mathcal{P}^2}}(t)$ over the partition $\mathcal{P}$ as in (6.16); therefore, the FTS of system (2.1) with respect to $(t_0, T, \mathcal{X}_{R_{\mathcal{P}^1}}, \mathcal{X}_{\Gamma_{\mathcal{P}^2}}(\cdot))$ is equivalent to the FTS with respect to $(t_0, T, \mathcal{X}_{\tilde{R}_{\mathcal{P}}}, \mathcal{X}_{\tilde{\Gamma}_{\mathcal{P}}}(\cdot))$.

Now, let us consider a candidate PQLF $P_{\mathcal{P}}(x, t)$ with x_0 such that $\tilde{R}_{\mathcal{P}}(x_0) \leq 1$; in view of (6.15b), we have

$$\tilde{\Gamma}_{\mathcal{P}} \left(x(t, x_0), t \right) \leq P_{\mathcal{P}} \left(x(t, x_0), t \right), \quad t \in [t_0, t_0 + T].$$

Since condition (6.15a) implies that $\dot{P}_{\mathcal{P}}(x, t)$ is negative definite along the trajectories of system (2.1), we have

$$P_{\mathcal{P}} \left(x(t, x_0), t \right) \leq P_{\mathcal{P}}(x_0, t_0), \quad t \in [t_0, t_0 + T].$$

Finally, in view of (6.15c),

$$P_{\mathcal{P}}(x_0, t_0) \leq \tilde{R}_{\mathcal{P}}(x_0) \leq 1.$$

We can conclude that $x_0 \in \mathcal{X}_{\tilde{R}_{\mathcal{P}}}$ implies $\tilde{\Gamma}_{\mathcal{P}}(x(t, x_0), t) < 1, t \in [t_0, t_0 + T]$, i e., $x(t, x_0) \in \mathcal{X}_{\tilde{\Gamma}_{\mathcal{P}}}(t), t \in [t_0, t_0 + T]$. ◇

Without loss of generality, in the following, we will refer to the case of domains defined over the same partition in order to simplify the notation. Therefore, in the sequel, we shall assume that $\mathcal{P}^1 = \mathcal{P}^2 =: \mathcal{P} = \{S_1, \ldots, S_r\}$.

Although important from a theoretical viewpoint, the conditions of Theorem 6.2 cannot be easily applied because they require to check three infinite-dimensional inequalities (see (6.15a)–(6.15c)) and the infinite-dimensional equality constraints (6.14).

The problem of reducing the infinite-dimensional inequalities (6.15a)–(6.15c) to DLMIs, through an S-procedure approach, follows the guidelines of [28] and will be discussed later in the chapter; therefore, the remainder of this section will focus on the space-continuity constraints (6.14) in order to resort to a set of conditions amenable of numerical computation. To this end, as a first step, for any possible candidate PQLF,we shall reduce conditions (6.14) to a finite number of equalities.

It is worth noting that these conditions will not introduce any conservativeness with respect to equations (6.14).

In order to state the next result, note that, given a conical partition $\mathcal{P} = \{S_1, \ldots, S_r\}$ of $\mathbb{R}^n$, the intersection between two members of the partition S_i and S_j, namely $H(S_i, S_j)$, either reduces to the zero vector, or, according to (6.4), we have

$$\text{extr}\big(H(S_i, S_j)\big) = \text{extr}(S_i) \cap \text{extr}(S_j). \tag{6.17}$$

In this last case, from (6.17) it readily follows that the extremal rays generating the intersection of S_i and S_j are a subset of the set of the generating rays of $\mathcal{P}$, namely $\mathcal{R}_\mathcal{P} = \{\hat{x}_1, \ldots, \hat{x}_v\}$.

Lemma 6.1 *Let the PQLF* $P_\mathcal{P}(x, t)$, $t \in [t_0, t_0 + T]$, *be defined over the conical partition* $\mathcal{P} = \{S_1, \ldots, S_r\}$ *of* $\mathbb{R}^n$. *Then* $P_\mathcal{P}(x, t)$ *is space-continuous if and only if*

$$\hat{x}_h^T P_i(t)\hat{x}_h = \hat{x}_h^T P_j(t)\hat{x}_h, \tag{6.18a}$$

$$\hat{x}_k^T P_i(t)\hat{x}_k = \hat{x}_k^T P_j(t)\hat{x}_k, \tag{6.18b}$$

$$\hat{x}_h^T P_i(t)\hat{x}_k = \hat{x}_h^T P_j(t)\hat{x}_k, \tag{6.18c}$$

for all pairs of vectors $\hat{x}_h, \hat{x}_k$ *taken from* $\text{extr}(H(S_i, S_j))$ *and for* $i, j = 1, \ldots, r$, $i \neq j$, $t \in [t_0, t_0 + T]$. $\square$

Remark 6.3 When, as it is usual in $\mathbb{R}^2$, for given i and j, the set $\text{extr}(H(S_i, S_j))$ reduces to one element, say $\hat{x}_h$, conditions (6.18a)–(6.18c) collapse into the single equality (6.18a). Obviously, conditions (6.18a)–(6.18c) are trivially satisfied when, for given i and j, the set $H(S_i, S_j)$ reduces to the zero vector. $\diamond$

Remark 6.4 It is important to point out the correct interpretation of Lemma 6.1 since this will be a key point for the following developments. The statement of the lemma says that the product $\hat{x}_h^T P_i(t)\hat{x}_k$ does depend on h, k and *does not* depend on i. To clarify this fact, assume, for instance, that for some $i, j \in \{1, \ldots, r\}$, $i \neq j$, we have

$$\text{extr}\big(H(S_i, S_j)\big) = \{\ldots, \hat{x}_h, \ldots, \hat{x}_k, \ldots\}$$

and that for some $l, m \in \{1, \ldots, r\}$, $l \neq m$, possibly different from i, j, we have

$$\text{extr}\big(H(S_l, S_m)\big) = \{\ldots, \hat{x}_h, \ldots, \hat{x}_k, \ldots\}.$$

Then the statement of Lemma 6.1 ensures that, for $t \in [t_0, t_0 + T]$,

$$\hat{x}_h^T P_i(t)\hat{x}_h = \hat{x}_h^T P_j(t)\hat{x}_h = \hat{x}_h^T P_l(t)\hat{x}_h = \hat{x}_h^T P_m(t)\hat{x}_h, \tag{6.19a}$$

$$\hat{x}_k^T P_i(t)\hat{x}_k = \hat{x}_k^T P_j(t)\hat{x}_k = \hat{x}_k^T P_l(t)\hat{x}_k = \hat{x}_k^T P_m(t)\hat{x}_k, \tag{6.19b}$$

$$\hat{x}_h^T P_i(t)\hat{x}_k = \hat{x}_h^T P_j(t)\hat{x}_k = \hat{x}_h^T P_l(t)\hat{x}_k = \hat{x}_h^T P_m(t)\hat{x}_k. \tag{6.19c}$$

$\diamond$

Proof The proof is divided into two parts. First, we prove that, for given i and j, conditions (6.18a)–(6.18c) are necessary and sufficient for the space-continuity of $P_{\mathcal{P}}(x, t)$ over $H(S_i, S_j)$; then we shall show that conditions (6.18a)–(6.18c) do not depend on the particular i and j considered.

Therefore, let us assume that i and j are given, and let $\text{extr}(H(S_i, S_j)) = \{\tilde{x}_1, \ldots, \tilde{x}_z\}$.[1]

Sufficient condition. A generic element $x \in H(S_i, S_j)$ can be expressed as

$$x = \sum_{h=1}^{z} \lambda_h \tilde{x}_h, \quad \lambda_h \geq 0.$$

Since the quadratic forms $x^T P_i(t) x$ and $x^T P_j(t) x$ with $x \in H(S_i, S_j)$ can be written as

$$x^T P_i(t) x = \left(\sum_{h=1}^{z} \lambda_h \tilde{x}_h \right)^T P_i(t) \left(\sum_{h=1}^{z} \lambda_h \tilde{x}_h \right)$$

$$= \sum_{h=1}^{z} \lambda_h^2 \left(\tilde{x}_h^T P_i(t) \tilde{x}_h \right) + 2 \sum_{k=1}^{z-1} \sum_{h=k+1}^{z} \lambda_k \lambda_h \left(\tilde{x}_k^T P_i(t) \tilde{x}_h \right), \quad (6.20)$$

$$x^T P_j(t) x = \left(\sum_{h=1}^{z} \lambda_h \tilde{x}_h \right)^T P_j(t) \left(\sum_{h=1}^{z} \lambda_h \tilde{x}_h \right)$$

$$= \sum_{h=1}^{z} \lambda_h^2 \left(\tilde{x}_h^T P_j(t) \tilde{x}_h \right) + 2 \sum_{k=1}^{z-1} \sum_{h=k+1}^{z} \lambda_k \lambda_h \left(\tilde{x}_k^T P_j(t) \tilde{x}_h \right), \quad (6.21)$$

it follows that conditions (6.18a)–(6.18c) ensure the continuity of the Lyapunov function over $H(S_i, S_j)$.

Necessary condition. Let us assume that the PQLF function $P_{\mathcal{P}}(x, t)$ is space-continuous over $H(S_i, S_j)$. Then

$$x^T P_i(t) x = x^T P_j(t) x, \quad x \in H(S_i, S_j).$$

Hence, for any pair $\tilde{x}_h, \tilde{x}_k$ taken from $\text{extr}(H(S_i, S_j)) \subset H(S_i, S_j)$, we have that

$$\tilde{x}_h^T P_i(t) \tilde{x}_h = \tilde{x}_h^T P_j(t) \tilde{x}_h, \quad (6.22a)$$

$$\tilde{x}_k^T P_i(t) \tilde{x}_k = \tilde{x}_k^T P_j(t) \tilde{x}_k. \quad (6.22b)$$

---

[1]For the sake of rigor, we use different symbols to denote the elements of the generic set $\text{extr}(H(S_i, S_j))$, namely $\tilde{x}$, with respect to the elements of the set $\mathcal{R}_{\mathcal{P}}$, namely $\hat{x}$; indeed, $\text{extr}(H(S_i, S_j))$ is a strict subset of $\mathcal{R}_{\mathcal{P}}$, and therefore the same element has to be indexed differently whether we consider it as a member of $\text{extr}(H(S_i, S_j))$ or a member of $\mathcal{R}_{\mathcal{P}}$.

Moreover, since $H(S_i, S_j)$ is a conical set, the vector $(\tilde{x}_h + \tilde{x}_k)$ belongs to $H(S_i, S_j)$, and hence

$$(\tilde{x}_h + \tilde{x}_k)^T P_i(t)(\tilde{x}_h + \tilde{x}_k) = (\tilde{x}_h + \tilde{x}_k)^T P_j(t)(\tilde{x}_h + \tilde{x}_k).$$

The last equality implies that

$$\tilde{x}_h^T P_i(t)\tilde{x}_h + \tilde{x}_k^T P_i(t)\tilde{x}_k + 2\tilde{x}_h^T P_i(t)\tilde{x}_k = \tilde{x}_h^T P_j(t)\tilde{x}_h + \tilde{x}_k^T P_j(t)\tilde{x}_k + 2\tilde{x}_h^T P_j(t)\tilde{x}_k. \tag{6.23}$$

Finally, because of conditions (6.22a)–(6.22b), the latter equation implies that

$$\tilde{x}_h^T P_i(t)\tilde{x}_k = \tilde{x}_h^T P_j(t)\tilde{x}_k.$$

To conclude the proof, we shall show that conditions (6.18a)–(6.18c) are independent on the particular i and j chosen. Assume that, for some $i, j \in \{1, \ldots, r\}$, $i \neq j$, we have

$$\text{extr}(H(S_i, S_j)) = \{\ldots, \hat{x}_h, \ldots, \hat{x}_k, \ldots\} \tag{6.24}$$

and that for some $l, m \in \{1, \ldots, r\}$, $l \neq m$, possibly different from i, j, we have

$$\text{extr}(H(S_l, S_m)) = \{\ldots, \hat{x}_h, \ldots, \hat{x}_k, \ldots\}. \tag{6.25}$$

We have already proven that, for the pair i, j, from (6.24) it follows

$$\hat{x}_h^T P_i(t)\hat{x}_h = \hat{x}_h^T P_j(t)\hat{x}_h, \tag{6.26a}$$

$$\hat{x}_k^T P_i(t)\hat{x}_k = \hat{x}_k^T P_j(t)\hat{x}_k, \tag{6.26b}$$

$$\hat{x}_h^T P_i(t)\hat{x}_k = \hat{x}_h^T P_j(t)\hat{x}_k. \tag{6.26c}$$

Using the same arguments, (6.25) implies that

$$\hat{x}_h^T P_l(t)\hat{x}_h \hat{x}_h^T P_m(t)\hat{x}_h, \tag{6.27a}$$

$$\hat{x}_k^T P_l(t)\hat{x}_k = \hat{x}_k^T P_m(t)\hat{x}_k, \tag{6.27b}$$

$$\hat{x}_h^T P_l(t)\hat{x}_k = \hat{x}_h^T P_m(t)\hat{x}_k. \tag{6.27c}$$

On the other hand, since the pair $\hat{x}_h, \hat{x}_k$ belongs to both $\text{extr}(H(S_i, S_j))$ and $\text{extr}(H(S_l, S_m))$, we can conclude that the pair $\hat{x}_h, \hat{x}_k$ also belongs to $\text{extr}(H(S_i, S_l))$; therefore, we can write

$$\hat{x}_h^T P_i(t)\hat{x}_h = \hat{x}_h^T P_l(t)\hat{x}_h, \tag{6.28a}$$

$$\hat{x}_k^T P_i(t)\hat{x}_k = \hat{x}_k^T P_l(t)\hat{x}_k, \tag{6.28b}$$

$$\hat{x}_h^T P_i(t)\hat{x}_k = \hat{x}_h^T P_l(t)\hat{x}_k. \tag{6.28c}$$

Putting together (6.26a)–(6.26c), (6.27a)–(6.27c) and (6.28a)–(6.28c), we get (6.19a)–(6.19c); this concludes the proof.   ◇

At this point, taking advantage of the above conditions and making use of $\mathcal{S}$-Procedure arguments [59], computationally tractable conditions to verify the FTS of system (2.1) with PQDs can be derived.

Theorem 6.3 [29] *Let us consider the LTV system* (2.1) *and two PQDs* $\mathcal{X}_{R_{\mathcal{P}}}$ *and* $\mathcal{X}_{\Gamma_{\mathcal{P}}}(t)$, $t \in [t_0, t_0 + T]$, *defined over the partition* $\mathcal{P} = \{S_1, \ldots, S_r\}$ *of* $\mathbb{R}^n$. *Then system* (2.1) *is finite-time stable with respect to* $(t_0, T, \mathcal{X}_{R_{\mathcal{P}}}, \mathcal{X}_{\Gamma_{\mathcal{P}}}(\cdot))$ *if there exists a PQLF* $P_{\mathcal{P}}(x, t)$, *such that*

(i) *the equality constraints for space-continuity* (6.18a)–(6.18c) *are satisfied;*
(ii) *given a set of matrices* $Q_{i,k} \in \mathbb{R}^{n \times n}$ *verifying*

$$x^T Q_{i,k} x \leq 0, \quad x \in S_i, \quad i = 1, \ldots, r, \quad k = 1, \ldots, s_i,$$

there exist positive scalars $b_{i,k}$ *and positive scalar functions* $c_{i,k}(t)$, $v_{i,k}(t)$ *satisfying the DLMI conditions*

$$\dot{P}_i(t) + A(t)^T P_i(t) + P_i(t) A(t) - \sum_{k=1}^{s_i} c_{i,k}(t) Q_{i,k} < 0, \qquad (6.29a)$$

$$P_i(t) - \Gamma_i(t) + \sum_{k=1}^{s_i} v_{i,k}(t) Q_{i,k} \geq 0, \qquad (6.29b)$$

$$P_i(t_0) - R_i - \sum_{k=1}^{s_i} b_{i,k} Q_{i,k} \leq 0, \qquad (6.29c)$$

for $i = 1, \ldots, r$, $k = 1, \ldots, s_i$, *and* $t \in [t_0, t_0 + T]$. □

Proof According to Lemma 6.1, condition (i) guarantees the space-continuity of the PQLF $P_{\mathcal{P}}(x, t)$. By exploiting $\mathcal{S}$-Procedure arguments, it readily follows that conditions (6.15a)–(6.15c) are ensured by (6.29a)–(6.29c), as shown in [59]. ◇

Note that Theorem 6.3 recasts the FTS problem, by using PQLFs, into a computationally tractable optimization problem since we now deal with a finite number of equalities and inequalities. Nevertheless, the new formulation may lead to numerical conditioning problems due to the presence of the equality constraints (6.18a)–(6.18c). To overcome this problem, in the following section we propose a reparameterization of the matrices $P_i(t)$, $i = 1, \ldots, r$, which allows us to get rid of the equality constraints without adding further conservativeness.

6.3.1 Reparameterization of the Quadratic Forms

In this section, the space-continuity conditions of Lemma 6.1 are stated in a form that enables a more efficient solution from the numerical point of view. Referring, as usual, to the partition $\mathcal{P}$ of $\mathbb{R}^n$ such that $\mathcal{R}_\mathcal{P} = \{\hat{x}_1, \ldots, \hat{x}_v\}$, we use the fact that, according to Remark 6.4, the quadratic form $\hat{x}_h^T P_i(t) \hat{x}_k$ does not depend on i. Therefore, we can define the symmetric matrix-valued function $\Phi(t) = \{\phi_{hk}(t)\} \in \mathbb{R}^{v \times v}$ such that

$$\phi_{hk}(t) := \hat{x}_h^T P_i(t) \hat{x}_k.$$

Without loss of generality, let us restrict our attention to PQLFs such that the set of normalized extremal rays of each cone S_i has cardinality n, i.e., $\text{extr}(S_i) = \{\hat{x}_{i_1}, \ldots, \hat{x}_{i_n}\} \subset \mathcal{R}_\mathcal{P}$.[2] Moreover, let Λ_i be the *selection matrix* that captures the matrix with the extremal rays of S_i, namely $\mathcal{M}_{\text{extr}(S_i)} = [\hat{x}_{i_1} \ldots \hat{x}_{i_n}]$, from the matrix containing all the generating rays of the conical partition $\mathcal{M}_{\mathcal{R}_\mathcal{P}} = [\hat{x}_1 \ldots \hat{x}_v]$, i.e.,

$$\mathcal{M}_{\text{extr}(S_i)} = \mathcal{M}_{\mathcal{R}_\mathcal{P}} \Lambda_i.$$

Now let us define the matrix function containing the parameters associated with the ith cone as

$$\Phi_i(t) = \begin{bmatrix} \phi_{i_1 i_1}(t) & \phi_{i_1 i_2}(t) & \ldots & \phi_{i_1 i_n}(t) \\ * & \phi_{i_2 i_2}(t) & \ldots & \ldots \\ * & * & \ldots & \phi_{i_{n-1} i_n}(t) \\ * & * & * & \phi_{i_n i_n}(t) \end{bmatrix}, \quad i = 1, \ldots, r. \quad (6.30)$$

Clearly, the matrix functions $\Phi(\cdot)$ and $\Phi_i(\cdot)$ are related by $\Phi_i(\cdot) = \Lambda_i^T \Phi(\cdot) \Lambda_i$. According to (6.30), the continuity conditions (6.18a)–(6.18c) for the ith cone may be rewritten as

$$\begin{bmatrix} \hat{x}_{i_1}^T \\ \vdots \\ \hat{x}_{i_n}^T \end{bmatrix} P_i(t) \begin{bmatrix} \hat{x}_{i_1} & \ldots & \hat{x}_{i_n} \end{bmatrix} = \Phi_i(t)$$

or, equivalently,

$$\mathcal{M}_{\text{extr}(S_i)}^T P_i(t) \mathcal{M}_{\text{extr}(S_i)} = \Phi_i(t).$$

By the definition of conical partition, each cone has dimension n, and then $\mathcal{M}_{\text{extr}(S_i)}$ is invertible. Hence,

$$P_i(t) = \mathcal{M}_{\text{extr}(S_i)}^{-T} \Phi_i(t) \mathcal{M}_{\text{extr}(S_i)}^{-1}$$

$$= (\mathcal{M}_{\mathcal{R}_\mathcal{P}} \Lambda_i)^{-T} \Lambda_i^T \Phi(t) \Lambda_i (\mathcal{M}_{\mathcal{R}_\mathcal{P}} \Lambda_i)^{-1}. \quad (6.31)$$

This last equation reparameterizes $P_i(t)$ as a function of $\Phi(t)$ without introducing any conservativeness. In this way, we are able to state a numerically efficient condition to test the FTS of system (2.1) with PQDs.

---

[2] As a matter of fact, any cone with more than n extremal rays can be partitioned into a collection of cones of cardinality n.

Theorem 6.4 (FTS of CT-LTV via PQLFs, [29]) *Let us consider the LTV system (2.1) and two PQDs $\mathcal{X}_{R_{\mathcal{P}}}$ and $\mathcal{X}_{\Gamma_{\mathcal{P}}}(t)$, $t \in [t_0, t_0 + T]$, defined over the partition $\mathcal{P} = \{S_1, \dots, S_r\}$ of $\mathbb{R}^n$. Then system (2.1) is finite-time stable with respect to $(t_0, T, \mathcal{X}_{R_{\mathcal{P}}}, \mathcal{X}_{\Gamma_{\mathcal{P}}}(\cdot))$ if, given a set of matrices $Q_{i,k} \in \mathbb{R}^{n \times n}$ such that*

$$x^T Q_{i,k} x \leq 0, \quad x \in S_i, \quad i = 1, \dots, r, \quad k = 1, \dots, s_i,$$

there exist positive scalars $b_{i,k}$, positive scalar functions $c_{i,k}(t)$, $v_{i,k}(t) > 0$, and a piecewise continuously differentiable matrix function $\Phi(t) \in \mathbb{R}^{v \times v}$ satisfying the DLMI conditions

$$\dot{P}_i(t) + A(t)^T P_i(t) + P_i(t)A(t) - \sum_{k=1}^{s_i} c_{i,k}(t) Q_{i,k} < 0, \tag{6.32a}$$

$$P_i(t) - \Gamma_i(t) + \sum_{k=1}^{s_i} v_{i,k}(t) Q_{i,k} \geq 0, \tag{6.32b}$$

$$P_i(t_0) - R_i - \sum_{k=1}^{s_i} b_{i,k} Q_{i,k} \leq 0, \tag{6.32c}$$

for $t \in [t_0, t_0+T]$, $i = 1, \dots, r, k = 1, \dots, s_i$, where the matrices $P_i(\cdot)$, $i = 1, \dots, r$, are defined as in (6.31). □

Remark 6.5 To reduce the conservatism of the $\mathcal{S}$-Procedure in Theorem 6.4, the quadratic forms $x^T Q_{i,k} x$, $i = 1, \dots, r$, $k = 1, \dots, s_i$, should be greater than zero outside the cone S_i. A possible strategy for the choice of the matrices $Q_{i,k}$ in the n-dimensional case will be presented below. For the sake of simplicity, we will drop the pedices i and k, and hence we will focus on a single cone S and a single matrix Q.

Let us consider a conical set $S = \text{cone}\{\hat{x}_1, \dots, \hat{x}_n\} \subset \mathbb{R}^n$. Given a set of scalars $\alpha_{j,h} \geq 0$, $j, h = 1, \dots, n$, with $j \neq h$ and $\alpha_{j,h} = \alpha_{h,j}$, a matrix Q such that $x^T Q x \leq 0$ for all $x \in S$ can be chosen as follows:

$$\hat{x}_j^T Q \hat{x}_j^T = 0, \quad j = 1, \dots, n, \tag{6.33}$$

$$\hat{x}_j^T Q \hat{x}_h^T = -\alpha_{j,h}, \tag{6.34}$$

for $j, h = 1, \dots, n$ and $j \neq h$.
It is worth noticing that

- conditions (6.33)–(6.34) assure that the quadratic form $x^T Q x$ is nonpositive inside the cone S and null on its generating rays;
- a matrix Q satisfying conditions (6.33)–(6.34) is univocally identified. Indeed, the unknowns of a symmetric matrix $Q \in \mathbb{R}^n$ and the number of constraints imposed by Eqs. (6.33)–(6.34) are $\frac{(n^2-n)}{2} + n$ (n constraints are imposed by Eq. (6.33) and $\binom{n}{2} = \frac{n!}{2!(n-2)!} = \frac{n(n-1)}{2}$ by Eq. (6.34)). ◊

To conclude this section, we present a corollary of Theorem 6.4. This corollary points out an important property of the level curves of the PQLF in the proposed FTS approach.

Corollary 6.1 *Let us assume that system (2.1) is finite-time stable with respect to $(t_0, T, \mathcal{X}_{R_{\mathcal{P}}}, \mathcal{X}_{\Gamma_{\mathcal{P}}})$ and that there exists a PQLF $P_{\mathcal{P}}(x, t)$ verifying the conditions of Theorem 6.4. Then, system (2.1) is finite-time stable with respect to $(t_0, \tilde{T}, \mathcal{X}_{R_{\mathcal{P}}}, \tilde{X}(\tilde{T}))$, where*

$$\tilde{X}(\tilde{T}) := \bigcup_{\tau \in [t_0, t_0 + \tilde{T}]} \mathcal{X}_{P_{\mathcal{P}}}(\tau) \subseteq \mathcal{X}_{\Gamma_{\mathcal{P}}} \tag{6.35}$$

and $\tilde{T} \in [0, T]$.

Proof Since $P_{\mathcal{P}}(x, t)$ verifies the conditions of Theorem 6.4, given $x_0 \in \mathcal{X}_{R_{\mathcal{P}}}$ and a time instant $\tau \in [t_0, t_0 + T]$, we have

$$x(\tau, x_0) \in \mathcal{X}_{P_{\mathcal{P}}}(\tau) \subseteq \mathcal{X}_{\Gamma_{\mathcal{P}}}. \tag{6.36}$$

Hence, given $x_0 \in \mathcal{X}_{R_{\mathcal{P}}}$, for all $t \in [t_0, t_0 + \tilde{T}]$,

$$x(t, x_0) \in \tilde{X}(\tilde{T}) = \bigcup_{\tau \in [t_0, t_0 + \tilde{T}]} \mathcal{X}_{P_{\mathcal{P}}}(\tau). \qquad \qquad \Diamond$$

6.4 Numerical Examples

The example section is divided into two parts. In the first part, we consider an FTS problem where the initial domain is assumed to be polytopic, while the trajectory domain is ellipsoidal. In the second part, an FTS problem with polytopic initial and trajectory domains is illustrated.

These examples remark the effectiveness of the PQLF-based approach, both for what concerns the possibility of considering domains belonging to different classes and with regards to the use of time-varying Lyapunov functions whose level curves can fit the different shape of the domains. Moreover, in these cases, the conditions of Theorem 2.1 can only be applied with added conservativeness due to the necessity of approximating polytopic domains by ellipsoidal domains.

6.4.1 Polytopic Initial Domain and Ellipsoidal Trajectory Domain

Let us consider the following CT-LTI system with all unstable evolution modes:

$$\dot{x} = \begin{pmatrix} 0.1 & 0 \\ 0 & 0.1 \end{pmatrix} x, \tag{6.37}$$

Fig. 6.2 The state trajectories of system (6.37) starting from the vertices of the initial polytopic domain for $\tilde{t} \in \{2.5, 5, 7.5, 10\}$ (*green lines*). *The red shapes* denote the initial and the trajectory domains for the FTS problem with respect to $(0, T, \mathcal{X}_{R_\mathcal{P}}, X(\gamma^*))$; *the blue shape denotes the trajectory domain defined by $\tilde{X}(\tilde{t})$ at the time instants $\tilde{t}$ listed above*

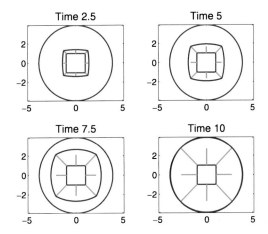

and a squared initial domain $\mathcal{X}_{R_\mathcal{P}}$, where the conical partition is given by $\mathcal{P} = \{S_1, S_2, S_3, S_4\}$ with

$$
\begin{aligned}
S_1 &= \mathrm{cone}\big(\{(1 \quad 1)^T ; (1 \quad -1)^T\}\big), \\
S_2 &= \mathrm{cone}\big(\{(1 \quad -1)^T ; (-1 \quad -1)^T\}\big), \\
S_3 &= \mathrm{cone}\big(\{(-1 \quad -1)^T ; (-1 \quad 1)^T\}\big), \\
S_4 &= \mathrm{cone}\big(\{(-1 \quad 1)^T ; (1 \quad 1)^T\}\big).
\end{aligned}
\tag{6.38}
$$

The PQLF $R_\mathcal{P}(x)$ is composed by the following matrices:

$$
\begin{aligned}
R_1 &= (1 \quad 0)^T (1 \quad 0), \\
R_2 &= (0 \quad -1)^T (0 \quad -1), \\
R_3 &= (-1 \quad 0)^T (-1 \quad 0), \\
R_4 &= (0 \quad 1)^T (0 \quad 1).
\end{aligned}
\tag{6.39}
$$

Finally, consider a circular time-invariant trajectory domain $X(\gamma)$ defined as

$$
X(\gamma) = \{x : x^T x < \gamma^2\}.
$$

With the aid of Theorem 6.4, we find that the maximum value of γ such that system (6.37) is finite-time stable with respect to $(0, T, \mathcal{X}_{R_\mathcal{P}}, X(\gamma^*))$, with $T = 10$, is $\gamma^* = 3.9223$ (see Fig. 6.2).

Note that, in this case, the conditions in Theorem 2.1 and the one contained in the main result of [49] can only be applied in a conservative way since they require the approximation of the initial domain by a suitable ellipsoid.

Finally, making use of Corollary 6.1, it is possible to show that system (6.37) is finite-time stable wrt $(0, \tilde{T}, \mathcal{X}_{R_\mathcal{P}}, \tilde{X}(\tilde{T}))$, where

$$
\tilde{X}(\tilde{T}) = \bigcup_{\tau \in [0, \tilde{T}]} \mathcal{X}_{P_\mathcal{P}}(\tilde{T}),
$$

Fig. 6.3 Mass-spring-friction
system

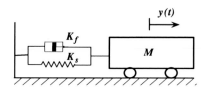

and $\tilde{T} \in [0, T]$. In Fig. 6.2, the trajectory domains defined by the Lyapunov function
at different time instants $\tilde{T} \in \{2.5, 5, 7.5, 10\}$ are shown. It is important to recognize
that these domains have initially a polytopic shape (as the initial domain $\mathcal{X}_{R_P}$), and
then they move toward an ellipsoidal shape (as the trajectory domain $X(\gamma^*)$).

6.4.2 Polytopic Initial and Trajectory Domains

6.4.2.1 Mass-Spring-Friction System

In this example, we consider the mass-spring-friction system in Fig. 6.3 described
by the following dynamical equation:

$$M\ddot{y}(t) + K_f \dot{y}(t) + K_s y(t) = 0, \qquad (6.40)$$

where y is the position of the mass expressed in meters, $M = 1$ Kg, $K_f = 0.25$ Ns/m, $K_s = 1$ N/m.

Let us assume that the following polytopic time-varying constraints are imposed
on the state variables:

$$-0.8 \le y(0) \le 0.8, \qquad (6.41a)$$

$$-2.5 \le \dot{y}(0) \le 2.5, \qquad (6.41b)$$

$$-0.8\alpha(1 - 0.5t/T) \le y(t) \le 0.8\alpha(1 - 0.5t/T), \quad t \in [0, T], \qquad (6.41c)$$

$$-2.5\alpha(1 - 0.5t/T) \le \dot{y}(t) \le 2.5\alpha(1 - 0.5t/T), \quad t \in [0, T], \qquad (6.41d)$$

where $\alpha \in \mathbb{R}$ and $T = 10$ s. The aim of this example is the evaluation of the minimum value of α such that system (6.40) is finite-time stable with respect to the
constraints (6.41a)–(6.41d).

In order to evaluate the effectiveness of the FTS conditions proposed in this chapter, we first applied the conditions of Theorem 2.1 with an appropriate ellipsoidal
approximation of the polytopic initial and trajectory domains. In this way, it is possible to prove that system (6.40) is finite-time stable with respect to the constraints
in (6.41a)–(6.41d) for $\alpha \ge 3.6$. This result is improved with the aid of Theorem 6.4
proving that system (6.40) is finite-time stable for $\alpha \ge 3.2$. In Figs. 6.4 and 6.5, the
free evolution of the state variables of system (6.40) from the vertices of the initial domain is illustrated. Moreover, the bound given by the time-varying trajectory
domain for $\alpha = 3.2$ and $\alpha = 3.6$ is shown.

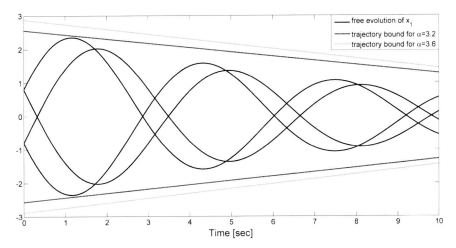

Fig. 6.4 Free evolution of $x_1(t)$ in (6.40) from the vertices of the initial domain

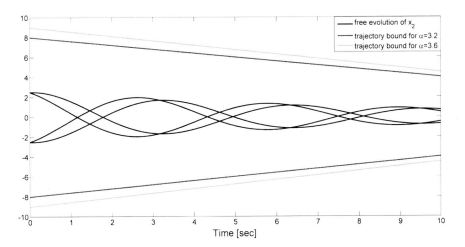

Fig. 6.5 Free evolution of $x_2(t)$ in (6.40) from the vertices of the initial domain

6.4.2.2 RLC Circuit

Let us consider the RLC circuit in Fig. 6.6 described by the dynamical equation

$$Ri_L(t) + v_C(t) + L\frac{d}{dt}i_L(t) = V, \qquad (6.42a)$$

$$C\frac{d}{dt}v_C(t) = i_L(t), \qquad (6.42b)$$

where i_L is the current in the inductor, v_C is the voltage across the capacitor, V is the voltage from the generator, $R = 0.25$ kΩ, $C = 10$ μF and $L = 10$ mH.

Fig. 6.6 RLC circuit

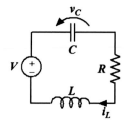

The aim of this example is the study of the transient dynamics of the circuit assuming that $V = 0$, while $v_C, i_L \neq 0$. In particular, let us assume the following constrains on i_L and v_C:

$$-2 \leq v_C(0) \leq 2, \tag{6.43a}$$

$$-0.2 \leq i_L(0) \leq 0.2, \tag{6.43b}$$

$$-2\alpha(1 - 0.9t/T) \leq v_C(t) \leq 2\alpha(1 - 0.9t/T), \quad t \in [0, T], \tag{6.43c}$$

$$-0.2\alpha(1 - 0.9t/T) \leq i_L(t) \leq 0.2\alpha(1 - 0.9t/T), \quad t \in [0, T], \tag{6.43d}$$

where $\alpha \in \mathbb{R}$ and $T = 0.015$ s.

Our goal is to evaluate the minimum value of α such that system (6.42a)–(6.42b) is finite-time stable with respect to the constraints (6.43a)–(6.43d).

The feasibility problem for system (6.42a)–(6.42b) with the constraints (6.43a)–(6.43d) can be easily recast into an FTS problem with polytopic constraints, where system (6.42a)–(6.42b) can be rewritten in the form (2.1) with

$$A = \begin{pmatrix} 0 & 10^5 \\ -100 & -2.5 \cdot 10^4 \end{pmatrix}, \quad x = \begin{pmatrix} x_1(t) \\ x_2(t) \end{pmatrix} := \begin{pmatrix} v_C(t) \\ i_L(t) \end{pmatrix}. \tag{6.44}$$

Applying the conditions of Theorem 2.1 with an appropriate ellipsoidal approximation of the polytopic initial and trajectory domains, it is possible to prove that system (6.44) is finite-time stable with respect to the constraints in (6.43a)–(6.43d) for $\alpha \geq 2.4$. Again, this result is improved with the aid of Theorem 6.4, proving that system (6.44) is finite-time stable for $\alpha \geq 1.5$.

In Figs. 6.7 and 6.8, the zero-input evolution of the state variables of system (6.44) from the vertices of the initial domain is illustrated. Moreover, the bound given by the time-varying trajectory domain for $\alpha = 1.5$ and $\alpha = 2.4$ is shown.

6.5 Summary

In this chapter, we have studied the FTS problem when the initial and trajectory domains are PQDs. A special class of Lyapunov functions, namely the class of PQLFs, whose level curves represent the boundaries of PQDs, has allowed us to derive a

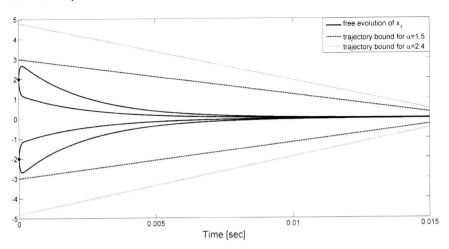

Fig. 6.7 Free evolution of $x_1(t)$ in (6.44) from the vertices of the initial domain

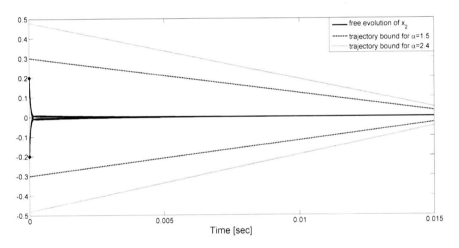

Fig. 6.8 Free evolution of $x_2(t)$ in (6.44) from the vertices of the initial domain

sufficient condition for FTS. Such a condition requires the solution of three infinite-dimensional inequalities and one infinite-dimensional equality, needed to guarantee the continuity of the PQLF. Therefore, most part of the chapter has been devoted to transform the main result into a computationally tractable optimization problem.

It is interesting to recall that another approach has been developed in the literature to deal with polytopic domains [13, 16]; such an approach makes use of the class of polyhedral Lyapunov functions.

Polyhedral Lyapunov functions are strictly related to the class of PQLFs considered in this chapter since both are *universal* from the point of view of classical LS (see [35, 58]).

The improvement obtained via the PQLF-based methodology with respect to the results in [13, 16] is twofold. First of all, when an FTS with polytopic initial and trajectory domains is dealt with, the PQLF approach leads to a DLMI-based optimization problem, while the main result of [16] required the solution of a nonconvex optimization problem. Therefore, it is expected that the approach developed in this chapter is less conservative than the methodology illustrated in [16]. Moreover, Theorem 6.4 allows us to deal with the more general case in which the initial and/or the trajectory domain are piecewise quadratic.

In order to illustrate the first point, let us consider again the mass-spring-friction system of Sect. 6.4.2.1, where the initial and trajectory domains are chosen as in [16], namely

$$-0.8 \leq y(0) \leq 0.8, \tag{6.45a}$$

$$-2.5 \leq \dot{y}(0) \leq 2.5, \tag{6.45b}$$

$$-2.4 \leq y(t) \leq 2.4, \quad t \in [0, T], \tag{6.45c}$$

$$-7.5 \leq \dot{y}(t) \leq 7.5, \quad t \in [0, T]. \tag{6.45d}$$

Let us denote by $\mathcal{X}_{R_{\mathcal{P}}}$ and $\mathcal{X}_{\Gamma_{\mathcal{P}}}$ the initial and trajectory domains, respectively, and by $\mathcal{P} = \{S_1, S_2, S_3, S_4\}$ the corresponding conical partition, where

$$S_1 = \text{cone}\big(\{(0.8 \quad 2.5)^T; (0.8 \quad -2.5)^T\}\big),$$

$$S_2 = \text{cone}\big(\{(0.8 \quad -2.5)^T; (-0.8 \quad -2.5)^T\}\big),$$

$$S_3 = \text{cone}\big(\{(-0.8 \quad -2.5)^T; (-0.8 \quad 2.5)^T\}\big),$$

$$S_4 = \text{cone}\big(\{(-0.8 \quad 2.5)^T; (0.8 \quad 2.5)^T\}\big).$$

The PQLF $R_{\mathcal{P}}(x)$ and $\Gamma_{\mathcal{P}}(x)$ are defined via the following diadic matrices:

$$R_1 = (1.25 \quad 0)^T (1.25 \quad 0), \qquad \Gamma_1 = \frac{1}{9} R_1,$$

$$R_2 = (0 \quad -0.4)^T (0 \quad -0.4), \qquad \Gamma_2 = \frac{1}{9} R_2,$$

$$R_3 = (-1.25 \quad 0)^T (-1.25 \quad 0), \qquad \Gamma_3 = \frac{1}{9} R_3,$$

$$R_4 = (0 \quad 0.4)^T (0 \quad 0.4), \qquad \Gamma_4 = \frac{1}{9} R_4.$$

In Fig. 6.9 the state trajectories starting from the vertices of the initial domain are shown. Figure 6.9 shows that if $x_0 \in \mathcal{X}_{R_{\mathcal{P}}}$, the trajectories of the system remain inside the domain $\mathcal{X}_{\Gamma_{\mathcal{P}}}$ for all $t \geq 0$, and they reach the minimum distance from the external boundary for $T \simeq 1$.

In [16], with the aid of the polyhedral functions approach, the authors obtained an estimate of the maximum allowable T, namely $T = 0.8$ s, such that the system is finite-time stable.

Fig. 6.9 The state
trajectories of system (6.40)
starting from the vertices of
the initial polytopic domain
(*green lines*). *The blue dots*
denote the values attained by
the state variables for $t = 1$ s;
the blue dashed lines divide
the state-space according to
the partition $\mathcal{P} = \{S_1, S_2,$
$S_3, S_4\}$; *the red lines* denote
the initial and the trajectory
domains for the FTS problem

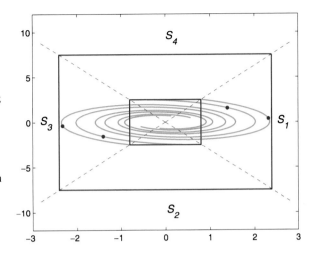

Then, the conditions in Theorem 6.4 have been applied to the present example.
In this way, it has been possible to prove that system (6.40) with the polytopic constraints in (6.45a)–(6.45d) is finite-time stable for $T \gg 1$; in particular, it is possible
to prove FTS of system (6.40) for $T = 10$ s by a piecewise linear matrix-valued
function $\Phi(\cdot)$. As expected, the obtained estimate of the maximum allowed T with
the PQLF approach strongly improves the result obtained in [16].

Note that, in this chapter, the Finite-Time Stabilization problem via PQLF is
not dealt with; this is due to the fact that, with the proposed machinery, the finite-
time stabilization conditions with PQLFs turn out to be differential *bilinear* matrix
inequalities, which cannot be converted into a convex optimization problem.

Part II
Hybrid Systems

Chapter 7
FTS of IDLSs

7.1 Introduction

In this paper, we consider the class of CT-LTV systems with finite state jumps, which are linear continuous-time systems whose states undergo finite jump discontinuities at discrete instants of time. Such systems can be regarded as a special class of hybrid systems, namely IDLSs [54], which can be either *time-dependent* if the state jumps are time-driven or *state-dependent* if the state jumps occur when the trajectory reaches an assigned subset of the state space, the so-called *resetting set*. An example which falls into this category of systems is the automatic gear-box in cruise control (for more details and further examples, see [76]).

Lyapunov stabilization of hybrid systems has been thoroughly discussed in the literature (see, for instance, the monographs [54, 67, 76] and references therein); obviously, this chapter is devoted to study the FTS problem for IDLSs.

To this regard, a necessary and sufficient condition for the FTS of TD-IDLSs will be provided; such a condition requires the solution of either a coupled differential-difference linear matrix inequality (D/DLMI) or a coupled difference/differential Lyapunov equation (D/DLE); concerning SD-IDLSs, a sufficient condition for FTS is derived. Differently from the TD-IDLS case, such a condition cannot be numerically implemented as a convex optimization problem; therefore, we provide a procedure that allows us to convert the sufficient condition into a numerically tractable problem (D/DLMIs) when the resetting set has a certain structure.

The chapter is ended by three examples. The first example shows how a sampled data system can be viewed as a TD-IDLS; then the FTS properties of such a system are studied by the theory developed in this chapter. The second (numerical) example deals with SD-IDLSs; finally, the third example considers the model, developed in the framework of TD-IDLSs, of a body that jumps on a two-dimensional elastic surface. Such an example will be continued in Chap. 9 to illustrate the synthesis of a feedback controller when the occurrence of the resetting times is uncertain.

F. Amato et al., *Finite-Time Stability and Control*,
Lecture Notes in Control and Information Sciences 453,
DOI 10.1007/978-1-4471-5664-2_7, © Springer-Verlag London 2014

7.2 IDLSs

7.2.1 State Space Representation

An IDLS is described by

$$\dot{x}(t) = A_c(t)x(t), \quad x(t_0) = x_0, \quad \big(t, x(t)\big) \notin \mathcal{S}, \tag{7.1a}$$

$$x^+(t) = A_d\big(t, x(t)\big)x(t), \quad \big(t, x(t)\big) \in \mathcal{S}, \tag{7.1b}$$

where $A_c(\cdot) : t \in \mathbb{R}_0^+ \mapsto \mathbb{R}^{n \times n}$ is a continuous matrix-valued function describing the *continuous-time* dynamics of the system, and $A_d(\cdot, \cdot) : (t, x(t)) \in \mathcal{S} \mapsto \mathbb{R}^{n \times n}$ is a matrix-valued function describing the *resetting law* of the system defined over the *resetting set* $\mathcal{S} \subset \mathbb{R}_0^+ \times \mathbb{R}^n$ (see [54]).

For a given trajectory $x(t)$ of the IDLS (7.1a)–(7.1b), let us denote by t_k, for $k \in \mathbb{N}$, the kth instant of time at which $(t, x(t))$ intersects $\mathcal{S}$ and call it *resetting time*. According to the continuous-time dynamics (7.1a) and the resetting law (7.1b), an IDLS presents a left-continuous trajectory with finite jump from $x(t_k)$ to $x^+(t_k) = \lim_{\epsilon \to 0^+} x(t_k + \epsilon)$ at each resetting time t_k.

Depending on the definition of the resetting set $\mathcal{S}$, IDLSs can be classified as follows [54]:

(i) TD-IDLSs. In this case, given a set $\mathcal{T} := \{t_1, t_2, \ldots\}$, $\mathcal{S}$ is defined as $\mathcal{S} = \mathcal{T} \times \mathcal{D}(x_0, \mathcal{T})$, where $\mathcal{D}(x_0, \mathcal{T}) = \{x(\bar{t}) : \bar{t} \in \mathcal{T}\} \subset \mathbb{R}^n$. For TD-IDLSs, the resetting set is defined by a prescribed sequence of time instants, which are independent of the state $x(\cdot)$, and we assume that the resetting law does not depend on the state value, that is,

$$x^+(t_k) = A_d(t_k)x(t_k), \quad t_k \in \mathcal{T}. \tag{7.2}$$

(ii) SD-ILDSs. In this case, given a set $\mathcal{D} \subset \mathbb{R}^n$, $\mathcal{S}$ is defined as $\mathcal{S} = \mathcal{T}(x_0, \mathcal{D}) \times \mathcal{D}$, where $\mathcal{T}(x_0, \mathcal{D}) = \{\bar{t} : x(\bar{t}) \in \mathcal{D}\} \subset \mathbb{R}_0^+$. For SD-IDLSs, the resetting set is defined by a region in the state space that does not depend on time. For SD-IDLSs, we assume that the resetting set is defined by the union of a finite number of regions $\mathcal{D}_j$ in the state space. Furthermore, the resetting law in each region is assumed to be constant, that is,

$$x^+(t) = A_{d,j}x(t), \quad x(t) \in \mathcal{D}_j, \quad j = 1, \ldots, N. \tag{7.3}$$

The notation $\mathcal{D}(x_0, \mathcal{T})$ makes clear the dependence of the resetting set on the initial state and on the set of resetting times in the case of TD-IDLSs; conversely, in the case of SD-IDLSs, $\mathcal{T}(x_0, \mathcal{D})$ indicates that the resetting times depend on the initial state and the resetting set.

Remark 7.1 When we consider a TD (SD)-IDLS, the set $\mathcal{D}(x_0, \mathcal{T})$ $(\mathcal{T}(x_0, \mathcal{D}))$ is not a priori known since it depends also on the initial state x_0.   ◊

Fig. 7.1 TD-IDLS as a
continuous-time linear system
with discrete jumps due to a
"Jump Generator System"

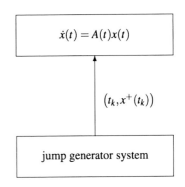

Remark 7.2 A TD-IDLS can be seen as a continuous-time system, as in (7.1a), whose state exhibits a finite jump according with the resetting law (7.1b) due to an external "jump generator system", as in Fig. 7.1. The scheme shown in Fig. 7.1 captures many cases of practical interest. For example, if $(t_k, x^+(t_k))$ is generated through an asynchronous impulsive input entering the system, we reobtain the class of impulsive control systems [66], whereas if $(t_k, x^+(t_k))$ is computed according to a given algorithm (for example, it is the output of a discrete event system [37]), the system depicted in Fig. 7.1 falls into the category of hybrid control systems [36]. ◊

The following two assumptions will be made when we deal with IDLSs. The first assures the well-posedness of the resetting times, while the second prevents system (7.1a)–(7.1b) from exhibiting Zeno behavior [61].

Assumption 7.1 *For all $t \in \mathbb{R}_0^+$ such that $(t, x(t)) \in \mathcal{S}$,*

$$\exists \varepsilon > 0 : \left(t + \delta, x(t + \delta)\right) \notin \mathcal{S} \quad \forall \delta \in \,]0, \varepsilon].$$ ◊

Assumption 7.2 *Given the interval $[t_0, t_0 + T]$, it includes only a finite number of resetting times. It follows that the resetting set to be considered in the time interval $[t_0, t_0 + T]$ is given by*

$$\mathcal{S} = \mathcal{T} \times \mathcal{D} \subset [t_0, t_0 + T] \times \mathbb{R}^n \quad \text{with } \mathcal{T} = \{t_1, t_2, \dots, t_r\}.$$ ◊

7.2.2 TD-IDLSs and TD-SLSs

The TD-IDLSs introduced in the previous section can also be seen as a special case of *time-dependent switching linear systems* (TD-SLSs, [24, 67]). This issue is briefly discussed in this section.

In order to define the class of TD-SLSs, the notion of *switching signal $\sigma(\cdot)$* is needed (see also [67]). Let $\sigma(\cdot) : \mathbb{R}_0^+ \mapsto \mathcal{P}$ be a piecewise constant function, where the discontinuities are the *resetting times*. The signal $\sigma(\cdot)$ is called a switching signal and is assumed to be right-continuous everywhere.

Let us now consider the family of linear systems

$$\dot{x}(t) = A_p(t)x(t), \tag{7.4}$$

where $p \in \mathcal{P} = \{1, \ldots, l\}$, and $A_p(\cdot) : \mathbb{R}_0^+ \mapsto \mathbb{R}^{n \times n}$ is a continuous matrix-valued function.

Given the family (7.4) and the switching signal $\sigma(\cdot)$, the class of TD-SLSs is given by

$$\dot{x}(t) = A_{\sigma(t)}(t)x(t), \quad x(t_0) = x_0, \quad t \notin \mathcal{T}, \tag{7.5a}$$

$$x(t_k^+) = A_d(t_k)x(t_k), \quad t_k \in \mathcal{T}. \tag{7.5b}$$

In particular, the switching signal $\sigma(t)$ specifies, at each time instant t, the linear system currently being *active*.

From the definition of TD-SLS given above it easily follows that the TD-IDLS (7.1a)–(7.1b) can also be seen as a TD-SLS when the special case of a single dynamic is considered. In Remark 7.5, we briefly discuss how to extend the results given to check FTS for TD-IDLSs to the case of TD-SLSs with different dynamics.

Furthermore, for the sake of simplicity and without loss of generality, the linear systems in the family (7.4) all have the same order. It should be noticed that after each resetting time, since the system matrix can change, the meaning of each state variable can change as well. This is particularly true in the case of systems with different orders. Further details on how to consider systems with different orders are given in Remark 7.6.

7.2.3 State Transition Matrix

In this section, the behavior of system (7.1a)–(7.1b) within a finite interval $[t_0, t_0 + T]$ is described. The solution of system (7.1a)–(7.1b) in such an interval is given by

$$x(t) = \Phi(t, t_0)x_0, \quad t \in [t_0, t_0 + T],$$

where the matrix function $\Phi(t, t_0)$ is the state transition matrix of system (7.1a)–(7.1b). The transition matrix turns out to be piecewise continuous with possible right discontinuities at the time instants t_k, $k = 1, \ldots, r$.

In the first interval, $\Phi(t, t_0)$ satisfies the following matrix equations:

$$\frac{\partial}{\partial t}\Phi(t, t_0) = A_c(t)\Phi(t, t_0), \quad t \in [t_0, t_1[, \quad \Phi(t_0, t_0) = I,$$

$$\Phi^+(t_1, t_0) = A_d(t_1, x(t_1))\Phi(t_1^-, t_0),$$

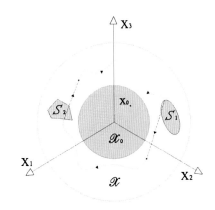

Fig. 7.2 Example of finite-time stable trajectory of an SD-IDLS. In this case, Γ is assumed time-invariant. The trajectory starts inside the sphere $\mathcal{X}_0$ defined by a positive definite matrix R and remains inside the sphere $\mathcal{X}$ defined by a constant positive definite matrix Γ for all $t \in [t_0, t_0 + T]$. When the trajectory reaches a resetting set, the state jumps

while in the following intervals, for $k = 1, \ldots, r - 1$, we have

$$\frac{\partial}{\partial t}\Phi(t, t_k) = A_c(t)\Phi(t, t_k), \quad t \in [t_k, t_{k+1}[, \quad \Phi(t_k, t_k) = \Phi^+(t_k, t_{k-1}),$$

$$\Phi^+(t_{k+1}, t_k) = A_d\big(t_k, x(t_k)\big)\Phi\big(t^-_{k+1}, t_k\big).$$

Finally, in the last interval, we have

$$\frac{\partial}{\partial t}\Phi(t, t_r) = A_c(t)\Phi(t, t_r), t \in [t_r, t_0 + T], \quad \Phi(t_r, t_r) = \Phi^+(t_r, t_{r-1}).$$

7.3 Problem Statement and Preliminaries

Definition 1.1 is here extended to the class of IDLSs where both the initial and trajectory domains are ellipsoids.

Definition 7.1 (FTS of IDLSs with Ellipsoidal Domains)   Given an initial time t_0, a positive scalar T, a positive definite matrix R, a positive definite matrix-valued function $\Gamma(\cdot)$ defined over $[t_0, t_0 + T]$, with $\Gamma(t_0) < R$, system (7.1a)–(7.1b) is said to be finite-time stable with respect to $(t_0, T, R, \Gamma(\cdot))$ if

$$x_0^T R x_0 \le 1 \Rightarrow x(t)^T \Gamma(t)x(t) < 1, \quad t \in [t_0, t_0 + T]. \tag{7.6}$$

$\diamondsuit$

As for the case of CT-LTVs, also Definition 7.1 can be interpreted in terms of ellipsoidal domains (see Fig. 7.2).

By following the guidelines of the proof of Theorem 2.1 we can prove the following necessary and sufficient condition for FTS of the IDLSs (7.1a)–(7.1b), based on the State Transition Matrix.

Theorem 7.1 [27]   *System* (7.1a)–(7.1b) *is finite-time stable with respect to* $(t_0, T, R, \Gamma(\cdot))$ *if and only if, for all* $t \in [t_0, t_0 + T]$,

$$\Phi^T(t, t_0)\Gamma(t)\Phi(t, t_0) < R, \tag{7.7}$$

where $\Phi(t, t_0)$ *is the state transition matrix of system* (7.1a)–(7.1b).   □

It is worth noticing that, for SD-IDLSs, condition (7.7) cannot be applied, since the resetting times are not *a priori* known and therefore $\Phi(\cdot, \cdot)$ cannot be computed. On the other hand, condition (7.7) may be difficult to check even if the resetting times are known, unless we are in the time-invariant case. This is due to the fact that the evaluation of the transition matrix for time-varying systems is a computationally hard problem. For these reasons, in the next section, we provide an alternative condition for FTS, which involves two coupled differential-difference Lyapunov inequalities.

7.4 FTS of TD-IDLSs

The main result of this section is the following theorem, which states two necessary and sufficient conditions for the FTS of TD-IDLSs. The proof exploits arguments similar to those of the proof of Theorem 2.1.

Theorem 7.2 (FTS of TD-IDLSs, [27])   *The following statements are equivalent:*

(i) *The TD-IDLS* (7.1a)–(7.1b) *is finite-time stable wrt* $(t_0, T, R, \Gamma(\cdot))$.
(ii) *The piecewise continuously differentiable matrix-valued solution* $W(\cdot)$:
     $[t_0, t_0 + T] \mapsto \mathbb{R}^{n \times n}$ *of the coupled D/DLE*

$$-\dot{W}(t) + A_c(t)W(t) + W(t)A_c^T(t) = 0, \quad t \in [t_0, t_0 + T], \quad t \notin \mathcal{T}, \tag{7.8a}$$

$$W^+(t_i) = A_d(t_i)W(t_i)A_d^T(t_i), \quad t_i \in \mathcal{T}, \tag{7.8b}$$

$$W(t_0) = R^{-1}, \tag{7.8c}$$

is positive definite and satisfies

$$C(t)W(t)C^T(t) < I, \quad t \in [t_0, t_0 + T], \tag{7.9}$$

where $C(\cdot)$ *is a nonsingular matrix-valued function such that* $\Gamma(t) = C^T(t)C(t)$ *in* $[t_0, t_0 + T]$.
(iii) *There exists a piecewise continuously differentiable symmetric matrix-valued function* $P(\cdot) : [t_0, t_0 + T] \mapsto \mathbb{R}^{n \times n}$ *that satisfies the following coupled D/DLMI:*

$$\dot{P}(t) + A_c^T(t)P(t) + P(t)A_c(t) < 0, \quad t \in [t_0, t_0 + T], \quad t \notin \mathcal{T}, \tag{7.10a}$$

$$P(t) > \Gamma(t), \quad t \in [t_0, t_0 + T], \tag{7.10b}$$

$$P(t_0) < R, \tag{7.10c}$$

$$A_d^T(t_i)P^+(t_i)A_d(t_i) - P(t_i) < 0, \quad t_i \in \mathcal{T}. \tag{7.10d}$$

□

Proof We will prove the equivalence of the three statements by showing first that (i) ⇔ (ii); then we show that (ii) ⇒ (iii) and that (iii) ⇒ (i).

(i) ⇔ (ii) Defining

$$W(t) = \Phi(t, t_0)R^{-1}\Phi^T(t, t_0) \tag{7.11}$$

and following the guidelines of the proof of the equivalence between conditions (i) and (iii) in Theorem 2.1, one can prove that $W(\cdot)$ satisfies (7.9) if and only if the TD-IDLS (7.1a)–(7.1b) is finite-time stable.

Now, in order to prove that $W(\cdot)$ is a solution of (7.8a)–(7.8c), let us consider the following properties of the transition matrix:

$$\Phi(t_0, t_0) = I, \tag{7.12a}$$

$$\dot{\Phi}(t, t_0) = A_c(t)\Phi(t, t_0), \quad t \notin \mathcal{T}, \tag{7.12b}$$

$$\Phi^+(t_i, t_{i-1}) = A_d(t_i)\Phi(t_i, t_{i-1}), \quad t_i \in \mathcal{T}. \tag{7.12c}$$

Hence, (7.8c) readily follows from (7.12a), while (7.8a) and (7.8b) can be obtained by computing $\dot{W}(t)$ and $W^+(t_i)$ from (7.11) and exploiting (7.12b) and (7.12c), respectively.

(ii) ⇒ (iii) We have already shown that if $W(\cdot)$ satisfies (7.8a)–(7.8c) and (7.9), then the TD-IDLS (7.1a)–(7.1b) is finite-time stable. Now, by continuity arguments, if (7.1a)–(7.1b) is finite-time stable, there exist two real scalars $\varepsilon_1, \varepsilon_2 > 0$ such that the following system also is finite-time stable:

$$\dot{z}(t) = \left(A_c(t) + \frac{\varepsilon_1}{2}\right)z(t), \quad z(t_0) = x_0, \quad t \notin \mathcal{T}, \tag{7.13a}$$

$$z^+(t_i) = A_d(t_i)\left(1 + \frac{\varepsilon_2}{2}\right)z(t_i), \quad t_i \in \mathcal{T}. \tag{7.13b}$$

Taking into account the equivalence of conditions (i) and (ii), we denote by $W_\varepsilon(\cdot)$ the piecewise continuously differentiable positive definite matrix-valued solution of

$$-\dot{W}_\varepsilon(t) + A_c(t)W_\varepsilon(t) + W_\varepsilon(t)A_c^T(t) + \varepsilon_1 W_\varepsilon(t) = 0,$$
$$t \in [t_0, t_0 + T], \, t \notin \mathcal{T}, \tag{7.14a}$$

$$W_\varepsilon^+(t_i) = A_d(t_i)W_\varepsilon(t_i)A_d^T(t_i) + \frac{\varepsilon_2^2}{4}A_d(t_i)W_\varepsilon(t_i)A_d^T(t_i)$$
$$+ \varepsilon_2 A_d(t_i)W_\varepsilon(t_i)A_d^T(t_i), \quad t_i \in \mathcal{T}, \tag{7.14b}$$

$$W_\varepsilon(t_0) = R^{-1}, \tag{7.14c}$$

which also satisfies $C(t)W_\varepsilon(t)C^T(t) < I$ with $t \in [t_0, t_0 + T]$. Note that the D/DLE (7.14a)–(7.14c) readily follows from (7.8a)–(7.8c) when we consider system (7.13a)–(7.13b). Exploiting again continuity arguments, it turns out that there exists a real scalar $\alpha > 1$ such that

$$\alpha C(t)W_\varepsilon(t)C^T(t) < I, \quad t \in [t_0, t_0 + T]. \tag{7.15}$$

Let $X(t) = \alpha W_\varepsilon(t)$ for all t in $[t_0, t_0 + T]$, inequality (7.15) reads

$$C(t)X(t)C^T(t) < I, \quad t \in [t_0, t_0 + T]. \tag{7.16}$$

Since $\dot{X}(t) = \alpha \dot{W}_\varepsilon(t)$, from (7.14a) we obtain

$$-\dot{X}(t) + A_c(t)X(t) + X(t)A_c^T(t) + \varepsilon_1 X(t) = 0$$

in $[t_0, t_0 + T]$ and for $t \notin \mathcal{T}$; taking into account that $X(t) > 0$, it follows that

$$-\dot{X}(t) + A_c(t)X(t) + X(t)A_c^T(t) < 0 \tag{7.17}$$

for all $t \notin \mathcal{T}$. Furthermore, from (7.14b) it readily follows that

$$X^+(t_i) > A_d(t_i)X(t_i)A_d^T(t_i), \tag{7.18}$$

while taking into account (7.14c), we obtain

$$X(t_0) > R^{-1}. \tag{7.19}$$

Finally, letting $P(t) = X^{-1}(t)$ for all t in $[t_0, t_0 + T]$ and applying the Schur complements to (7.18), inequalities (7.10a)–(7.10d) can be easily obtained from (7.16)–(7.19).

$\boxed{\text{(iii)} \Rightarrow \text{(i)}}$ Let $V(t, x) = x^T P(t)x$. Then, if $t \notin \mathcal{T}$, the derivative of V along the trajectories of system (7.1a) is

$$\dot{V}(t, x) = x^T \left(\dot{P}(t) + A_c(t)^T P(t) + P(t)A_c(t) \right)x,$$

which is negative definite by virtue of (7.10a). At the discontinuity points $t_k \in \mathcal{T}$, we have

$$V^+(t_k, x) - V(t_k, x) = x^{+T}(t_k)P^+(t_k)x^+(t_k) - x^T(t_k)P(t_k)x(t_k)$$
$$= x^T(t_k)\left(A_d(t_k)P^+(t_k)A_d(t_k) - P(t_k)\right)x(t_k),$$

which is negative semidefinite in view of (7.10b). We can conclude that $V(t, x)$ is strictly decreasing along the trajectories of system (7.1a)–(7.1b); hence, given x_0 such that $x_0^T Rx_0 \leq 1$, we have

$$x(t)^T \Gamma(t)x(t) \leq x(t)^T P(t)x(t) \quad \text{by (7.10c)}$$
$$< x_0^T P(t_0)x_0$$
$$< x_0^T Rx_0 \leq 1 \quad \text{by (7.10d).} \qquad \diamond$$

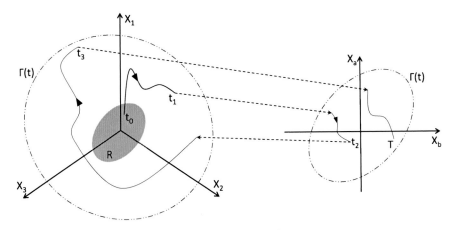

Fig. 7.3 Example of finite-time stable trajectory of a TD-SLS with three resetting time instants. In this case, $\Gamma(t)$ is assumed time-invariant between two resetting times

Remark 7.3 Similarly to what has been shown in Sect. 2.4 for CT-LTV systems, also for TD-IDLSs, the condition based on the coupled D/DLE in Theorem 7.2 turns out to be much more efficient from the computational point of view (see also Sect. 7.6). However, as in the CT-LTV case, the D/DLMI-based approach can be used to derive design conditions, as it will be shown in Chap. 8, while the D/DLE condition is not useful for such a purpose.   ◇

Remark 7.4 In general, the matrix-valued function $P(\cdot)$ in Theorem 7.2 is assumed piecewise continuous with finite jumps in the resetting times $t_k, k \in \mathbb{N}$. This structure for $P(\cdot)$ can be realized only if the jump instants are known and hence in the case of TD-IDLSs. In particular, condition (iii) in Theorem 7.2 reduces the FTS analysis problem of TD-IDLSs to a feasibility problem in the matrix variable $P(\cdot)$. When the structure of the matrix $P(\cdot)$ is fixed a priori, for example, piecewise affine (see the example in Sect. 7.7), the feasibility D/DLMI problem can be turned into a classical optimization problem involving LMIs [38].   ◇

Remark 7.5 The D/DLE and D/DLMI conditions for the FTS of TD-IDLSs given in Theorem 7.2 clearly hold also for the TD-SLSs introduced in Sect. 7.2.2. Indeed, since the two matrix-valued functions $W(\cdot)$ and $P(\cdot)$ are assumed to be piecewise continuously differentiable, conditions (ii) and (iii) hold also if the time-varying continuous-time dynamics (7.1a) have discontinuities in correspondence of the resetting times. Therefore, considering $A_c(\cdot)$ discontinuous when $t \in \mathcal{T}$ is equivalent to considering a TD-SLS.   ◇

Remark 7.6 Theorem 7.2 holds also for TD-SLSs when the family (7.4) is composed of systems with different dimensions. In this case, the single matrix-valued function $\Gamma(\cdot)$ should be replaced by $r+1$ matrix functions[1] $\Gamma_j(\cdot)$, $j = 1, \ldots, r+1$, defined on the subintervals $[t_{j-1}, t_j]$, $t_j \in \mathcal{T}$. These matrix-valued functions are square but do not necessarily have the same dimension. An example of a finite-time stable trajectory for a TD-SLS with different dynamic dimension is shown in Fig. 7.3, where the state trajectory starts inside the ellipsoid defined by the matrix R in the three-dimensional space and is limited by a three-dimensional weighting matrix $\Gamma_1(\cdot)$. After the first resetting time t_1, the system becomes of the second order, and a two-dimensional $\Gamma_2(\cdot)$ is accordingly defined.

Also, when dealing with this more general definition of TD-SLSs, the D/DLE (D/DLMI) of Theorem 7.2 can still be written by choosing $r+1$ matrix-valued functions $W_j(\cdot)$ and $(P_j(\cdot))$ with the same dimension of $A_{\sigma(t_j)}(\cdot)$ for all t and by noticing that, in this case, $A_d(t_k)$ is not a square matrix.   ◊

7.5 FTS of SD-IDLSs

The two necessary and sufficient conditions introduced in Theorem 7.2 can be applied only if the resetting times are a priori known, which is the case when TD-IDLSs are considered. Similar conditions can be applied also to SD-IDLSs; however, they become only sufficient to check FTS.

Indeed, it is possible to slightly modify the D/DLMI feasibility problem presented in condition (iii) of Theorem 7.2 in order to apply it as a sufficient condition to SD-IDLSs. In particular, if the two inequalities (7.10a) and (7.10d) hold for all $t \in [t_0, t_0 + T]$, then the given SD-IDLS is finite-time stable wrt $(t_0, T, R, \Gamma(\cdot))$.

It turns out that if the resetting times are not priori known, the matrix-valued function $P(\cdot)$ cannot exhibit jumps (see Remark 7.4), and the satisfaction of (7.10d) in $[t_0, t_0 + T]$ implies the Lyapunov stability of the resetting law (7.1b) for all t_k, leading to a sufficient condition for FTS.

When SD-IDLSs are considered, a less conservative sufficient condition to check FTS, which does not require the Lyapunov stability of the resetting law, is given by the following theorem.

Theorem 7.3 (FTS of SD-IDLSs, [28]) *Assume that the coupled differential-difference Lyapunov inequalities with terminal and initial conditions*

$$\dot{P}(t) + A_c(t)^T P(t) + P(t)A_c(t) < 0, \quad t \in (t_0, t_0 + T], \tag{7.20a}$$

$$x^T\left(A_{d,j}^T P(t)A_{d,j} - P(t)\right)x \leq 0, \quad t \in (t_0, t_0 + T], \quad x \in \mathcal{D}_j, \quad j = 1, \ldots, N, \tag{7.20b}$$

$$P(t) > \Gamma(t), \quad t \in [t_0, t_0 + T], \tag{7.20c}$$

$$P(t_0) < R, \tag{7.20d}$$

---

[1] r is the number of resetting times in $[t_0, t_0 + T]$, i.e., is the cardinality of $\mathcal{T}$.

admit a continuously differentiable symmetric solution $P(\cdot)$. *Then the SD-IDLS* *(7.1a)–(7.1b) is finite-time stable with respect to* $(t_0, T, R, \Gamma(\cdot))$.   □

Condition (7.20b) of Theorem 7.3 needs to be verified just on the resetting sets $\mathcal{D}_j$; hence, this formulation is not of the LMI form. In the following section, exploiting the S-procedure, a new D/DLMI condition will be derived for the finite-time stability analysis of SD-IDLSs.

7.5.1 Preliminary Results on Quadratic Forms

As will be shown later, the main result of the section requires to check whether, given a connected closed set $\mathcal{D} \subseteq \mathbb{R}^n$ and a symmetric matrix $Q_0 \in \mathbb{R}^{n \times n}$, the inequality

$$x^T Q_0 x < 0, \quad x \in \mathcal{D} \setminus \{0\}, \tag{7.21}$$

is satisfied.

In the following, our goal is to find some numerically tractable conditions that guarantee the satisfaction of (7.21). Exploiting S-procedure arguments [38, p. 24], it is readily seen that Q_0 satisfies (7.21) if the following feasibility problem admits a solution.

Problem 7.1 Given a connected closed set $\mathcal{D} \subset \mathbb{R}^n$, a symmetric matrix $Q_0 \in \mathbb{R}^{n \times n}$, and symmetric matrices $Q_i \in \mathbb{R}^{n \times n}$ satisfying

$$x^T Q_i x \leq 0, \quad x \in \mathcal{D}, \quad i = 1, \ldots, p, \tag{7.22a}$$

find nonnegative scalars c_i, $i = 1, \ldots, p$, such that

$$Q_0 - \sum_{i=1}^{p} c_i Q_i < 0. \tag{7.22b}$$

◊

The usefulness of Problem 7.1 relies on the fact that it can be recast in the LMI framework, where the coefficients c_i are the optimization variables of the LMI (7.22b). Clearly, a method to choose the matrices Q_i is needed. In the following, we also provide a procedure to build a suitable set of matrices Q_i, which can be exploited when the set $\mathcal{D}$ satisfies some assumptions.

As mentioned above, if Problem 7.1 admits a feasible solution, then (7.21) is satisfied. In general, the converse is not true. Therefore, it makes sense to investigate under which conditions solving Problem 7.1 is equivalent to check condition (7.21); the answer is given by the following lemmas.

Given a set $\mathcal{S} \subseteq \mathbb{R}^n$, in this chapter, we define the conical hull of $\mathcal{S}$ as

$$\text{cone}(\mathcal{S}) := \left\{ \lambda_1 x_1 + \cdots + \lambda_k x_k : \{x_1, \ldots, x_k\} \subseteq \mathcal{S}, \ \lambda_i \geq 0 \right\}.$$

Lemma 7.1 *Consider a nonempty connected closed set $\mathcal{D} \subseteq \mathbb{R}^n$ and a symmetric matrix $Q_0 \in \mathbb{R}^{n \times n}$. Then (7.21) is satisfied if and only if*

$$x^T Q_0 x < 0, \quad x \in \left(\text{cone}(\mathcal{D}) \cup \text{cone}(-\mathcal{D})\right) \setminus \{0\}. \tag{7.23}$$

□

Proof (7.23) $\Rightarrow$ (7.21). Trivial, since $\mathcal{D} \subseteq \text{cone}(\mathcal{D}) \cup \text{cone}(-\mathcal{D})$.
   (7.21) $\Rightarrow$ (7.23). Any point $\bar{x} \in \text{cone}(\mathcal{D}) \setminus \{0\}$ can be written as

$$\bar{x} = \lambda_1 x_1 + \cdots + \lambda_k x_k,$$

where $\{x_1, \ldots, x_k\} \subseteq \mathcal{D}$, and $\lambda_i \geq 0$. Since $\mathcal{D}$ is a connected set, it follows that there exist a scalar $\tilde{\lambda} \geq 0$ and a vector $\tilde{x} \in \mathcal{D}$ such that $\bar{x} = \tilde{\lambda}\tilde{x}$. Therefore,

$$\bar{x}^T Q_0 \bar{x} = \tilde{\lambda}^2 \tilde{x}^T Q_0 \tilde{x} < 0.$$

A similar statement can be made for every point $\bar{x} \in \text{cone}(-\mathcal{D}) \setminus \{0\}$. The implication easily follows.   ◇

Lemma 7.2 *Given a connected closed set $\mathcal{D} \subset \mathbb{R}^n$ and a symmetric matrix $Q_0 \in \mathbb{R}^{n \times n}$, assume that there exists a symmetric matrix $\bar{Q} \in \mathbb{R}^{n \times n}$ such that*

$$x^T \bar{Q} x \leq 0, \quad x \in \left(\text{cone}(\mathcal{D}) \cup \text{cone}(-\mathcal{D})\right) \setminus \{0\}, \tag{7.24a}$$

$$x^T \bar{Q} x > 0, \quad x \in \mathbb{R}^n \setminus \left(\text{cone}(\mathcal{D}) \cup \text{cone}(-\mathcal{D})\right), \tag{7.24b}$$

$$\exists \tilde{x} \in \mathcal{D} \setminus \{0\} : \tilde{x}^T \bar{Q} \tilde{x} < 0. \tag{7.24c}$$

Then condition (7.21) is equivalent to the feasibility Problem 7.1 with $p = 1$ and $Q_1 = \bar{Q}$.   □

Proof The proof is trivial once it is recognized that:

1. $x^T Q_0 x < 0$ for all $x \in \mathcal{D}$ if and only if $x^T Q_0 x < 0$ for all $x \in \text{cone}(\mathcal{D}) \cup \text{cone}(-\mathcal{D})$, by Lemma 7.1.
2. solving Problem 7.1 with $p = 1$ and $Q_1 = \bar{Q}$ is equivalent to applying the lossless S-procedure [28, 59].   ◇

   In Sect. 7.5.3, it will be shown that when the set $\mathcal{D}$ satisfies certain assumptions, the hypotheses of Lemma 7.2 are fulfilled, and the approach via Problem 7.1 does not add conservatism in the FTS analysis.

7.5.2 D/DLMI Condition for the FTS of SD-IDLSs

Let us consider the sufficient condition for FTS of the SD-IDLS (7.1a)–(7.1b) stated in Theorem 7.3. For a given t, condition (7.20b) is equal to (7.21) if we let $Q_0 =$

$A_{d,j}^T P(t) A_{d,j} - P(t)$ and $\mathcal{D} = D_j$ for $j = 1, \ldots, N$. Therefore, by exploiting the machinery introduced in Sect. 7.5.1, we can relax inequality (7.20b) and replace it with (see Problem 7.1)

$$A_{d,j}^T P(t) A_{d,j} - P(t) - \sum_{i=1}^{p_j} c_{i,j}(t) Q_{i,j} < 0, \quad i = 1, \ldots, p_j, \quad j = 1, \ldots, N,$$

where $Q_{i,j}$ are given symmetric matrices satisfying $x^T Q_{i,j} x \leq 0$ for all x in $\mathcal{D}_j$, and $c_{i,j}(t) \geq 0$ for $i = 1, \ldots, p_j$, $j = 1, \ldots, N$, and $t \in [t_0, t_0 + T]$.

On the basis of this consideration, we can immediately derive the following theorem.

Theorem 7.4 [28]   *Given a set of symmetric matrices* $Q_{i,j}$, $i = 1, \ldots, p_j$, $j = 1, \ldots, N$, *satisfying*

$$x^T Q_{i,j} x \leq 0, \quad x \in \mathcal{D}_j, \quad i = 1, \ldots, p_j, \quad j = 1, \ldots, N, \tag{7.25}$$

assume that there exist a continuously differentiable symmetric matrix function $P(\cdot)$ *and nonnegative scalar functions* $c_{i,j}(\cdot)$, $i = 1, \ldots, p_j$, $j = 1, \ldots, N$, *such that, for all* $t \in [t_0, t_0 + T]$,

$$\dot{P}(t) + A(t)^T P(t) + P(t) A(t) < 0, \tag{7.26a}$$

$$A_{d,j}^T P(t) A_{d,j} - P(t) - \sum_{i=1}^{p_j} c_{i,j}(t) Q_{i,j} < 0, \quad i = 1, \ldots, p_j, \quad j = 1, \ldots, N,$$
$$\tag{7.26b}$$

$$P(t) \geq \Gamma(t), \tag{7.26c}$$

$$P(t_0) < R. \tag{7.26d}$$

Then the SD-IDLS (7.1a)–(7.1b) *is finite-time stable wrt* $(t_0, T, R, \Gamma(\cdot))$. $\square$

Remark 7.7   In view of the results given in Theorem 7.4, it is now possible to clarify the usefulness of the formulation introduced in Problem 7.1. Such a formulation, indeed, allows us to replace condition (7.20b) with condition (7.26b). Note that, in principle, the former requires to solve an infinite number of time-varying inequalities over the sets $\mathcal{D}_k$, while the latter is just a set of LMIs that can be easily solved in an efficient way.   ◇

7.5.3   Analysis of Some Cases of Interest

Theorem 7.4 may introduce conservativeness with respect to Theorem 7.3 since, in general, the S-procedure is lossy. However, if for every resetting set $\mathcal{D}$ there exists a symmetric matrix Q that satisfies conditions (7.24a)–(7.24c), then Theorem 7.4 is

Fig. 7.4 Construction of the
set $\mathcal{P}_H(\mathcal{D})$

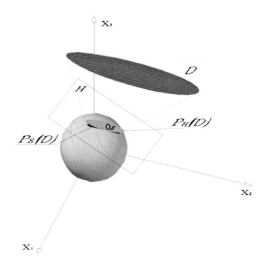

equivalent to Theorem 7.3. In this section, we will discuss two cases where Theorem 7.4 does not introduce conservatism: resetting sets in $\mathbb{R}^2$ and *ellipsoidal* resetting sets, we prove that the conservatism can be eliminated in both cases, except for the following trivial cases:

1. $\mathcal{D} \subseteq \mathbb{R}^n$ lies on a hyperplane that intersects the origin;
2. $\mathcal{D} \subseteq \mathbb{R}^n$ has dimension less than $n - 1$.

The definition of an ellipsoidal resetting set is based on the following constructive geometrical procedure.

Definition 7.2 (Projection wrt the Origin) Consider a hypersurface $H \subset \mathbb{R}^n$ and a set $\mathcal{D} \subseteq \mathbb{R}^n$. The projection of $\mathcal{D}$ on H with respect to the origin is defined as

$$\mathcal{P}_H(\mathcal{D}) := \{y \in H : y = \lambda x, \lambda \in \mathbb{R}, x \in \mathcal{D}\}. \qquad \qquad \diamond$$

Procedure 7.1 (Construction of $\mathcal{P}_H(\mathcal{D})$) Given a connected closed set $\mathcal{D} \subset \mathbb{R}^n$, construct the set $\mathcal{P}_H(\mathcal{D})$ as follows:

1. denote by $\mathcal{P}_s(\mathcal{D})$ the projection, with respect to the origin, of $\mathcal{D}$ on the unit sphere $s = \{x \in \mathbb{R}^n : x^T x = 1\}$;
2. denote by 0_s the Chebyshev center of $\mathcal{P}_s(\mathcal{D})$;[2]
3. denote by H the hyperplane of dimension $n - 1$ orthogonal to the line that joins the origin to 0_s and such that $0_s \in H$;
4. $\mathcal{P}_H(\mathcal{D})$ is the projection, with respect to the origin, of $\mathcal{D}$ on the hyperplane H.

An example of construction of the set $\mathcal{P}_H(\mathcal{D})$ is shown in Fig. 7.4. $\qquad \diamond$

---

[2] The Chebyshev center of a set $\mathcal{D} \subseteq \mathbb{R}^n$ is defined as $0_{\mathcal{D}} := \arg\min_{x \in \mathbb{R}^n} (\max_{\theta \in S} \|x - \theta\|_\infty)$.

Definition 7.3 (Ellipsoidal Resetting Set)   Consider a *nontrivial* resetting set $\mathcal{D} \subset \mathbb{R}^n$ and construct the set $\mathcal{P}_H(\mathcal{D})$ using Procedure 7.1. If $\mathcal{P}_H(\mathcal{D})$ is a hyper-ellipsoid of dimension $n - 1$, then $\mathcal{D}$ is called an *ellipsoidal resetting set*.   ◇

Remark 7.8   Since $\mathcal{P}_H(\mathcal{D})$ is constructed using two projections with respect to the origin, it follows that $\mathrm{cone}(\mathcal{D}) = \mathrm{cone}(\mathcal{P}_H(\mathcal{D}))$.   ◇

The following theorem provides a necessary and sufficient condition that enables to find a symmetric matrix $Q \in \mathbb{R}^{2 \times 2}$ satisfying conditions (7.24a)–(7.24c).

Theorem 7.5 [28]   *Every* nontrivial *resetting set* $\mathcal{D}$ *in* $\mathbb{R}^2$ *admits a symmetric matrix* $Q \in \mathbb{R}^{2 \times 2}$ *that satisfies conditions* (7.24a)–(7.24c).   □

Proof   To prove our statement, we provide a procedure to calculate a matrix Q satisfying conditions (7.24a)–(7.24c). Let $s_1, s_2 \in \mathcal{D}$ be such that, denoting $\bar{\mathcal{D}} = \mathrm{conv}(\{s_1, s_2\})$, we have $\mathrm{cone}(\bar{\mathcal{D}}) = \mathrm{cone}(\mathcal{D})$. Then, taking into account Lemma 7.2, condition (7.24a)–(7.24c) can be equivalently evaluated on the set $\bar{\mathcal{D}}$. In particular, considering the properties of the quadratic forms, it is easy to verify that such a condition can be replaced by the following condition:

$$x^T Q x < 0, \quad x \in \mathrm{int}(\bar{\mathcal{D}}), \tag{7.27a}$$

$$x^T Q x = 0, \quad \text{for } x = s_1, s_2, \tag{7.27b}$$

$$x^T Q x > 0, \quad x \in H \setminus \bar{\mathcal{D}}, \tag{7.27c}$$

where H is the hyperplane on which $\bar{\mathcal{D}}$ lies, and $\mathrm{int}(\bar{\mathcal{D}})$ denotes the interior of the set $\bar{\mathcal{D}}$. Letting $s_m = \frac{s_1 + s_2}{2}$, a symmetric matrix $Q \in R^{2 \times 2}$ such that $s_1^T Q s_1 = 0$, $s_2^T Q s_2 = 0$, and $s_m^T Q s_m < 0$ satisfies conditions (7.27a)–(7.27c).   ◇

The following theorem provides a sufficient condition to find a matrix $Q \in \mathbb{R}^{n \times n}$ that satsifies conditions (7.24a)–(7.24c).

Theorem 7.6 [28]   *If* $\mathcal{D}$ *is an ellipsoidal resetting set, then there exists a matrix* $Q \in \mathbb{R}^{n \times n}$ *that satisfies conditions* (7.24a)–(7.24c).   □

Proof   If $\mathcal{D}$ is an ellipsoidal resetting set, then $\mathrm{cone}(\mathcal{D}) = \mathrm{cone}(\mathcal{P}_H(\mathcal{D}))$ (see Remark 7.8). Taking into account Lemma 7.2, it follows that conditions (7.24a)–(7.24c) can be equivalently evaluated on the set $\mathcal{P}_H(\mathcal{D})$. In particular, considering the properties of the quadratic forms, it is easy to verify that such conditions can be replaced by the following condition:

$$x^T Q x < 0, \quad x \in \mathrm{int}(\mathcal{P}_H(\mathcal{D})), \tag{7.28a}$$

$$x^T Q x = 0, \quad x \in \partial \mathcal{P}_H(\mathcal{D}), \tag{7.28b}$$

$$x^T Q x > 0, \quad x \in H \setminus \mathcal{P}_H(\mathcal{D}). \tag{7.28c}$$

To conclude the proof, we need to show that the assumption of an ellipsoidal set $\mathcal{P}_H(\mathcal{D})$ is sufficient to find a matrix Q that satisfies conditions (7.28a)–(7.28c). In the sequel of the proof, we assume that:

- 0_s is on the nth coordinate axis, i.e.,

$$0_s = (0 \quad \ldots \quad 0 \quad r)^T, \quad r \in \mathbb{R}.$$

- The hyperplane H is orthogonal to the nth coordinate axis.

As a matter of fact, it is always possible, by means of opportune rotations, to satisfy these assumptions. In view of the assumptions made above, it is possible to describe the set $\partial \mathcal{P}_H(\mathcal{D})$ by the two equations

$$\frac{x_1^2}{a_1^2} + \cdots + \frac{x_{n-1}^2}{a_{n-1}^2} = 1, \quad x_n = r,$$

where $a_i \geq 0$, $i = 1, \ldots, n-1$. It is then straightforward to check that the matrix $Q = \mathrm{diag}(\frac{1}{a_1^2} \ldots \frac{1}{a_{n-1}^2} - \frac{1}{r^2})$ satisfies conditions (7.28a)–(7.28c).   ◇

7.6 Computational Issues

Let us consider the CT-LTV system (2.18) together with the following resetting law:

$$x^+(t) = \begin{pmatrix} 1.1 & 0 \\ 0 & 1.1 \end{pmatrix} x(t). \tag{7.29}$$

By letting

$$\mathcal{T} = \{0.1, 0.2, 0.3, 0.4, 0.5, 0.6, 0.7, 0.8, 0.9\}$$

and exploiting Theorem 7.2 it is possible to search for $\gamma_{\max}$ such that the TD-IDLS (2.18)–(7.29) is finite-time stable wrt $(0, 1, R, \Gamma)$, where the matrices R and Γ are given in (2.19).

When TD-IDLSs are considered, in order to recast the D/DLMI (7.10a)–(7.10d) in terms of LMIs, the matrix-valued function $P(\cdot)$ is assumed piecewise linear with jumps in correspondence of the resetting times (see Sect. 7.7.1).

Table 7.1 shows the results that have been obtained by using the same setup considered in Sect. 2.4. Comparing the results in Table 7.1 with that reported in Table 2.1, it should be also noticed that, when $P(\cdot)$ is piecewise linear with jumps, the LMI feasibility problem is solved more efficiently since it exploits additional degrees of freedom.

7.7 Examples

This section presents three examples where the analysis results introduced in this chapter are applied to both engineering and numerical systems.

Table 7.1 Values of γ_{max} satisfying Theorem 7.2 for the TD-IDLS system (2.18)–(7.29)

Condition	Sample time (T_s) [s]	γ_{max}	Average computation time for a single iteration [s]
DLMI (7.10a)–(7.10d)	0.1	0.3458	1.2
	0.05	0.3475	1.8
	0.025	0.3484	4.1
	0.005	0.3489	158.1
Solution of (7.8a)–(7.8c) and check of inequality (7.9)	$2 \cdot 10^{-4}$	0.4388	1.1

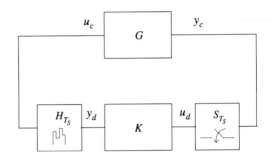

Fig. 7.5 Sampled-data system

7.7.1 Sampled-Data Systems as TD-IDLSs

In this example, we show how a sampled-data system can be expressed as a TD-IDLS.

Consider the sampled-data feedback system in Fig. 7.5, where G is a linear continuous-time, time-invariant plant described by

$$\dot{x}_c = A_c x_c + B_c u_c, \tag{7.30a}$$

$$y_c = C_c x_c, \tag{7.30b}$$

where $x_c \in \mathbb{R}^n$, $u_c \in \mathbb{R}^m$, $y_c \in \mathbb{R}^o$, $x_c(0) = x_{c0}$, and K is a linear discrete-time, time-invariant controller

$$x_d(k+1) = A_K x_d(k) + B_K u_d(k), \tag{7.31a}$$

$$y_d(k) = C_K x_d(k) + D_K u_d(k), \tag{7.31b}$$

where $x_d \in \mathbb{R}^p$, $u_d \in \mathbb{R}^o$, $y_d \in \mathbb{R}^m$, $x_d(0) = 0$.

The block labeled S_{T_s} represents the sampling operator with period T_s such that

$$S_{T_s} : y_c \rightarrow (S_{T_s} y_c) : (S_{T_s} y_c)(k) = y_c(kT_s), \tag{7.32}$$

and the block denoted by H_{T_s} represents the zero-order hold operator with time period T_s defined as

$$H_{T_s} : y_d \rightarrow (H_{T_s} y_d) : (H_{T_s} y_d)(t) = y_d(k), \quad kT_s < t \le (k+1)T_s. \tag{7.33}$$

Note that

$$u_d(k) = y_c(kT_s) = C_c x_c(kT_s), \tag{7.34a}$$

$$u_c(t) = y_d(k), \quad kT_s < t \le (k+1)T_s. \tag{7.34b}$$

Such an interconnection can be reduced to the following state space representation:

$$\dot{x}(t) = A(t)x(t), \quad x(0) = x_0, \quad t \ge 0, \ t \ne kT_s, \quad k = 1, 2, \dots, \tag{7.35a}$$
$$x^+(kT_s) = A_d(t_k)x(kT_s), \quad k = 1, 2, \dots.$$

By letting

$$x_1(\cdot) = x_c(\cdot),$$
$$x_2(t) = y_d(k), \quad kT_s < t \le (k+1)T_s,$$
$$x_3(t) = x_d(k+1), \quad kT_s < t \le (k+1)T_s,$$
$$x(0) = (x_{c0} \quad 0 \quad 0)^T,$$

and

$$A = \begin{pmatrix} A_c & B_c & 0 \\ 0 & 0 & 0 \\ 0 & 0 & 0 \end{pmatrix}, \quad A_d = \begin{pmatrix} I & 0 & 0 \\ D_K C_c & 0 & C_K \\ B_K C_c & 0 & A_K \end{pmatrix},$$

the sampled-data system can be expressed as a linear system with finite jumps where $t_k = kT_s$.

Remark 7.9 If we consider a sampled-data static state feedback system, i.e.,

$$y_d(k) = D_K u_d(k)$$

with $u_d(k) = x_c(kT_s)$, the state variables of the corresponding TD-IDLS (7.35a) reduce to

$$x_1(\cdot) = x_c(\cdot),$$
$$x_2(t) = y_d(k), \quad kT_s < t \le (k+1)T_s,$$

and

$$A = \begin{pmatrix} A_c & B_c \\ 0 & 0 \end{pmatrix}, \quad A_d = \begin{pmatrix} I & 0 \\ D_K & 0 \end{pmatrix}. \qquad \Diamond$$

Now, let us consider the sampled-data static state feedback system made up of the continuous-time linear plant G, defined by the matrices

$$A_c = \begin{pmatrix} 0 & 1 \\ -15 & -0.2 \end{pmatrix}, \qquad B_c = \begin{pmatrix} 0 \\ 1 \end{pmatrix}, \qquad C_c = \begin{pmatrix} 1 & 0 \\ 0 & 1 \end{pmatrix}, \qquad (7.36)$$

and the state feedback controller

$$D_K = -(0.0333 \quad 0.8519), \qquad (7.37)$$

designed through LQ optimization with $T_s = 0.015$ s. Condition (iii) of Theorem 7.2 is exploited to check the FTS of the closed-loop system with respect to $(0, T, R, \Gamma)$ when $T = 1$ s and

$$\Gamma = \mathrm{diag}(0.1, 0.003, 0.02), \qquad R = \mathrm{diag}(0.3, 0.02, 50).$$

Similarly to what has been done in Sects. 2.4 and 3.4, in order to turn the D/DLMI problem of condition (iii) into a standard LMI feasibility problem, the structure of the matrix-valued function $P(\cdot)$ is assumed piecewise affine with discrete jumps in correspondence of the known resetting times. In particular, we have

$$\begin{cases} P(0) = \Pi_1^0, \\ P(t) = \Pi_k^0 + \Pi_k^s (t - (k-1)T_s), \\ \quad k \in \mathbb{N}^+ : k \leq \bar{k}, \quad t \in](k-1)T_s, kT_s], \\ P(t) = \Pi_{\bar{k}+1}^0 + \Pi_{\bar{k}+1}^s (t - \bar{k}T_s), \quad t \in]\bar{k}T_s, T], \end{cases}$$

where $\bar{k} = \max\{k \in \mathbb{N}^+ : k < T/T_s\}$.

Given this structure for the matrix-valued function $P(\cdot)$, the time derivative $\dot{P}(\cdot)$ in (7.10a) is

$$\begin{cases} \dot{P}(t) = \Pi_k^s, \quad k \in \mathbb{N}^+ : k \leq \bar{k}, \quad t \in](k-1)T_s, kT_s], \\ \dot{P}(t) = \Pi_{\bar{k}+1}^s, \quad t \in]\bar{k}T_s, T], \end{cases}$$

while $P^+(t_k)$ in (7.10d) is given by

$$\begin{cases} P^+(t_k) = \Pi_k^0, \quad k \in \mathbb{N}^+ : k \leq \bar{k}, \quad t \in](k-1)T_s, kT_s], \\ P^+(t_k) = \Pi_{\bar{k}+1}^0, \quad t \in]\bar{k}T_s, T]; \end{cases}$$

hence, conditions (7.10a)–(7.10d) are reduced to a set of LMIs.

Using an optimization tool as the Matlab LMI Toolbox [48] together with the YALMIP parser [70], it is possible to find a set of matrices Π_k^s and Π_k^0, $k = 1, 2, \ldots, r$, such that conditions (iii) of Theorem 7.2 are satisfied. Therefore, we can conclude that the sampled-data system with matrices (7.36)–(7.37) is finite-time stable with respect to $(0, T, R, \Gamma)$.

7.7.2 FTS of SD-IDLSs

In this section, we consider the second-order SD-IDLS defined by the following matrices:

$$A = \begin{pmatrix} 0 & 1 \\ -1 + 0.3 \sin 10t & 0.5 \end{pmatrix},$$

$$A_{d,1} = \begin{pmatrix} 1.2 & 0 \\ 0 & -0.75 \end{pmatrix}, \qquad A_{d,2} = \begin{pmatrix} -0.72 & 0.16 \\ 0.13 & -0.78 \end{pmatrix},$$

where the two resetting sets $\mathcal{D}_1$ and $\mathcal{D}_2$ are given by

$$\mathcal{D}_1 = \mathrm{conv}\left(\begin{pmatrix} -0.4 \\ 0.6 \end{pmatrix}, \begin{pmatrix} 0.8 \\ 0.7 \end{pmatrix} \right),$$

$$\mathcal{D}_2 = \mathrm{conv}\left(\begin{pmatrix} -0.8 \\ -0.3 \end{pmatrix}, \begin{pmatrix} 0.4 \\ -0.6 \end{pmatrix} \right).$$

Note that the continuous dynamics is unstable, while the discrete dynamics described by $A_{d,1}$ is not Schur stable. We want to analyze the FTS of such an impulsive system for $t_0 = 5$ s, $T = 2.5$ s, and

$$\Gamma = \begin{pmatrix} 0.20 & -0.10 \\ -0.10 & 0.18 \end{pmatrix}, \qquad R = \begin{pmatrix} 7.5 & 4.5 \\ 4.5 & 9.0 \end{pmatrix}.$$

As usual, in order to recast the conditions provided in Theorem 7.4 in terms of LMIs, the matrix function $P(\cdot)$ is assumed to be piecewise linear. Theorem 7.5 assures that, for each of the considered resetting sets in $\mathbb{R}^2$, there exists a matrix Q that satisfies (7.24a)–(7.24c). Applying the procedure proposed in the proof of Theorem 7.5, the following matrices have been found:

$$Q_1 = \begin{pmatrix} -0.2909 & 0.0693 \\ 0.0693 & 0.2216 \end{pmatrix}, \qquad Q_2 = \begin{pmatrix} -0.2000 & 0.2000 \\ 0.2000 & 0.3556 \end{pmatrix}.$$

Using the matrices Q_1 and Q_2, it is possible to find a piecewise linear matrix $P(\cdot)$ and two nonnegative scalar functions $c_i(\cdot), i = 1, 2$, that satisfy conditions in Theorem 7.4. Hence, the considered system is finite-time stable with respect to (t_0, T, R, Γ). Now, let us change the two resetting laws as follows:

$$\hat{A}_{d,1} = \begin{pmatrix} -0.8 & 0 \\ 0 & 3.0 \end{pmatrix}, \qquad \hat{A}_{d,2} = \begin{pmatrix} 0.5 & 0 \\ 0 & 0.5 \end{pmatrix}.$$

It turns out that the considered SD-IDLS is not finite-time stable wrt (t_0, T, R, Γ), as shown by the trajectory in Fig. 7.6. Moreover, in this case it is not possible to satisfy the conditions in Theorem 7.4.

Fig. 7.6 Trajectory that is not finite-time stable wrt (t_0, T, R, Γ). The two resetting laws are given by $\hat{A}_{d,1}$ and $\hat{A}_{d,2}$, respectively

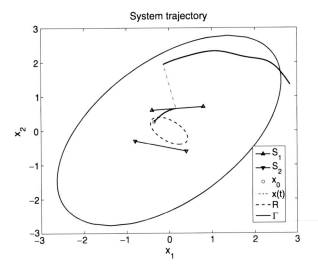

Fig. 7.7 The jumping body system considered in Sect. 7.7.3

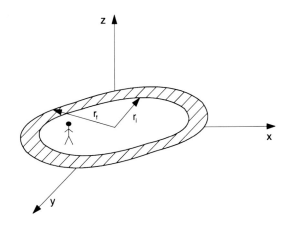

7.7.3 Jumping Body

In this section, we consider the dynamics of a body of mass $m = 30$ Kg, which jumps on a two-dimensional elastic surface. We assume that the elastic surface is circular with radius $r_f = 9$ m and that the body starts from a circular subregion of radius $r_i = 7$ m (see Fig. 7.7). Our aim is to determine the maximum time interval $\Omega = [0, \mathcal{N}]$, $\mathcal{N} \in \mathbb{N}$, such that the body is guaranteed to remain within the elastic surface. In order to derive a simplified TD-IDLS model of the considered system, we assume that the body impacts the surface at a frequency $f = 0.5$ Hz and that the vertical impact velocity v_z is constant. Indicating by x_1, x_3 the position of the body along the x and y axes, respectively, and by x_2, x_4 the velocity of the body along the x

and y axes, the model reads

$$
\begin{aligned}
\dot{x}_1(t) &= x_2(t), \\
\dot{x}_2(t) &= -\frac{c}{m}x_2(t), \\
\dot{x}_3(t) &= x_4(t), \qquad & \text{for } t \neq \frac{i}{f}, \ i = 1, 2, \ldots, \\
\dot{x}_4(t) &= -\frac{c}{m}x_4(t), \\
\dot{x}_5(t) &= 0,
\end{aligned}
$$

$$
\begin{aligned}
x_1^+(t_i) &= x_1(t_i), \\
x_2^+(t_i) &= \zeta x_2(t_i), \\
x_3^+(t_i) &= x_3(t_i), \qquad & \text{for } t_i = \frac{i}{f}, \ i = 1, 2, \ldots, \\
x_4^+(t_i) &= \zeta x_4(t_i), \\
x_5^+(t_i) &= x_5(t_i),
\end{aligned}
$$

where $c = 0.1$ is the air friction coefficient, and $\zeta = 0.9$ is the damping coefficient; $x_5(\cdot)$ is a fictitious state variable representing the vertical impact velocity v_z, which is assumed to be constant and equal to 1 m/s (more details about the model derivation can be found in [21]). The state variable $x_5(\cdot)$ will be exploited in Chap. 9 to generalize this example to the uncertain case.

FTS analysis is then carried out in order to find the maximum $\mathcal{N} \in \mathbb{N}$ such that the body does not exit the elastic surface during the time interval $\Omega = [0, \mathcal{N}]$. To this aim, the weighting matrices are selected as follows:

$$
R = \begin{pmatrix} \frac{1}{r_i^2} & 0 & 0 & 0 & 0 \\ 0 & 1 & 0 & 0 & 0 \\ 0 & 0 & \frac{1}{r_i^2} & 0 & 0 \\ 0 & 0 & 0 & 1 & 0 \\ 0 & 0 & 0 & 0 & \frac{1}{v_z^2} \end{pmatrix}, \qquad
\Gamma = \begin{pmatrix} \frac{1}{r_f^2} & 0 & 0 & 0 & 0 \\ 0 & 10^{-4} & 0 & 0 & 0 \\ 0 & 0 & \frac{1}{r_f^2} & 0 & 0 \\ 0 & 0 & 0 & 10^{-4} & 0 \\ 0 & 0 & 0 & 0 & 10^{-4} \end{pmatrix}.
$$

Note that, with the given choice of Γ, no constraint is assigned to the velocity of the body (see the elements equal to 10^{-4} in Γ). Furthermore, we assume that the body starts with a nonzero velocity only along the vertical axis (see the elements equal to 1 in R). Exploiting condition (ii) in Theorem 7.2, it turns out that the maximum value of $\mathcal{N}$ such that the considered system is finite-time stable wrt $(0, \mathcal{N}, R, \Gamma)$ is equal to 7.

7.8 Summary

In this chapter, the FTS of IDLSs has been considered. To this regard, the first result appeared in the literature was the sufficient condition for the FTS of TD-IDLSs provided in [15]; such a sufficient condition required the solution of a coupled D/DLMI.

TD-IDLSs are the most common and intuitive representation of IDLSs; however, it requires the a priori knowledge of the occurrence of the jump instants.

Often, in practical engineering problems, state jumps are activated when the state trajectory reaches a given set in the state space, the so-called resetting set; this is the case of SD-IDLSs. A sufficient condition for the FTS of such a class of systems has been published in [28] (see Theorem 7.6); however, differently from the TD-IDLS case, such a condition cannot be immediately converted into a convex optimization problem. Convex optimization results, derived by S-procedure arguments, can be obtained when the resetting set satisfies certain assumptions.

Finally, a necessary and sufficient condition for FTS of TD-IDLSs is given in Theorem 7.2; such a condition requires the solution of either a coupled D/DLMI or a coupled D/DLE. This result has been recently published in [27].

Some examples illustrate the application of Theorems 7.2 and 7.4–7.6, which deals with SD-IDLSs. In particular, the example of the jumping body will be continued in Chap. 9 to design a feedback controller in the uncertain resetting times case.

It is worth noting that the research on the FTS of impulsive and switching systems is currently rather active; see, for example, [69, 82, 83] in the context of linear systems (continuous and discrete-time), [80] in the area of stochastic Markovian systems, [86] in the framework of singular systems, and [84, 87] in the context of nonlinear systems.

Chapter 8
Controller Design for the Finite-Time Stabilization of IDLSs

8.1 Introduction

In this chapter, the finite-time stabilization problem for IDLSs is considered. The necessary and sufficient condition (iii) in Theorem 7.2 allows us to derive a necessary and sufficient condition for finite-time stabilization via output feedback of TD-IDLSs; as usual, such a condition requires the existence of a feasible solution to a coupled D/DLMI. The solution of the state feedback problem can be obtained as a special case of the general output feedback case.

Concerning SD-IDLSs, when the controller design is dealt with, differently from what has been done in Sect. 7.5 by using the S-procedure, even in the state-feedback case, the synthesis problem turns out to be a bilinear matrix inequalities feasibility problem [81], which is known to be a nonconvex optimization problem. In order to recast the FTS design of SD-IDLSs into a numerically tractable problem, it is required, at the price of some conservativeness, that the D/DLMI condition be satisfied for all $t \in [t_0, t_0 + T]$, rather than at the resetting times only.

In the example section at the end of the chapter, two examples are considered. The first (numerical) example deals with the SD-IDLS case, while, in the second example, which considers a classical engineering system composed of three interconnected reservoirs, the theory developed for TD-IDLSs is applied.

8.2 Problem Statement

In this section, the main problem dealt with in this chapter is precisely stated. The definition applies to both TD and SD-IDLSs; however, the two contexts require different approaches to solve the design problem.

Problem 8.1 (Finite-Time Stabilization of IDLSs) Consider the following IDLS:

$$\dot{x}(t) = A_c(t)x(t) + B(t)u(t), \quad x(t_0) = x_0, \quad (t, x(t)) \notin \mathcal{S}, \qquad (8.1a)$$

F. Amato et al., *Finite-Time Stability and Control*,
Lecture Notes in Control and Information Sciences 453,
DOI 10.1007/978-1-4471-5664-2_8, © Springer-Verlag London 2014

$$x^+(t_k) = A_d(t_k)x(t_k), \quad (t_k, x(t_k)) \in \mathcal{S}, \tag{8.1b}$$

$$y(t) = C(t)x(t) + D(t)u(t), \quad t > 0, \tag{8.1c}$$

where $u(t)$ is the control input, and $y(t)$ is the output. Given a positive number T, two positive definite matrices R and R_K, two positive definite matrix-valued functions $\Gamma(\cdot)$ and $\Gamma_K(\cdot)$ defined over $[t_0, t_0 + T]$, with $\Gamma(t_0) < R$, $\Gamma_K(t_0) < R_K$, find a dynamical output-feedback controller of the form

$$\dot{x}_K(t) = A_K(t)x_K(t) + B_K(t)y(t), \quad (t, x(t)) \notin \mathcal{S}, \tag{8.2a}$$

$$x_K^+(t_k) = A_{d,K}(t_k)x_K(t_k) + B_{d,K}(t_k)y(t_k), \quad (t_k, x(t_k)) \in \mathcal{S}, \tag{8.2b}$$

$$u(t) = C_K(t)x_K(t) + D_K(t)y(t), \tag{8.2c}$$

where $x_K(t)$ has the same dimension of $x(t)$, such that the closed-loop system obtained by the interconnection of (8.1a)–(8.1c) and (8.2a)–(8.2c) is finite-time stable with respect to $(t_0, T, \mathrm{diag}(R, R_K), \mathrm{diag}(\Gamma(\cdot), \Gamma_K(\cdot)))$.   ◇

Note that the dynamical controller (8.2a)–(8.2c) includes as special case the state-feedback controller. Indeed, when the state of system (8.1a)–(8.1c) is fully available, we can look for a state feedback finite-time stabilizing controller of the form $u(t) = K(t)x(t)$.

8.3 Finite-Time Stabilization of TD-IDLSs

Exploiting the D/DLMI feasibility problem introduced in condition (iii) of Theorem 7.2 (see also Remark 7.3), in this section, we derive necessary and sufficient conditions for finite-time stabilization of TD-IDLSs via output and state-feedback.

The sufficiency part of the next theorem has been proven in [22]; the fact that the condition is also necessary is proven here for the first time. It is a natural consequence of condition (iii) of Theorem 7.2.

Theorem 8.1 (Finite-Time Stabilization of TD-IDLSs) *Problem* 8.1 *is solvable for TD-IDLSs if and only if there exist piecewise continuously differentiable symmetric matrix-valued functions* $Q(\cdot)$ *and* $S(\cdot)$, *a nonsingular continuous matrix-valued function* $N(\cdot)$, *and continuous matrix-valued functions* $\hat{A}_K(\cdot)$, $\hat{B}_K(\cdot)$, $\hat{C}_K(\cdot)$, $D_K(\cdot)$, $\hat{A}_{d,K}(\cdot)$, *and* $\hat{B}_{d,K}(\cdot)$ *such that*

$$\begin{pmatrix} \Theta_{11} & \Theta_{12} \\ \Theta_{12}^T & \Theta_{22} \end{pmatrix} < 0, \quad t \in [t_0, t_0 + T], \quad t \notin \mathcal{T}, \tag{8.3a}$$

$$\begin{pmatrix} \Theta_{d,11} & \Theta_{d,12} \\ \Theta_{d,12}^T & \Theta_{d,22} \end{pmatrix} < 0, \quad t_k \in \mathcal{T}, \tag{8.3b}$$

$$
\begin{pmatrix}
Q & \Psi_{12} & \Psi_{13} & \Psi_{14} \\
\Psi_{12}^T & \Psi_{22} & 0 & 0 \\
\Psi_{13}^T & 0 & I & 0 \\
\Psi_{14}^T & 0 & 0 & I
\end{pmatrix} > 0, \quad t \in [t_0, t_0 + T], \tag{8.3c}
$$

$$
\begin{pmatrix} Q(t_0) & I \\ I & S(t_0) \end{pmatrix} < \begin{pmatrix} \Delta_{11} & Q(t_0)R \\ RQ(t_0) & R \end{pmatrix}, \tag{8.3d}
$$

where [1]

$$
\Theta_{11} = -\dot{Q} + A_c Q + Q A_c^T + B \hat{C}_K + \hat{C}_K^T B^T, \tag{8.4a}
$$

$$
\Theta_{12} = A_c + \hat{A}_K^T + B D_K C, \tag{8.4b}
$$

$$
\Theta_{22} = \dot{S} + S A_c + A_c^T S + \hat{B}_K C + C^T \hat{B}_K^T, \tag{8.4c}
$$

$$
\Theta_{d,11} = -\begin{pmatrix} Q & I \\ I & S \end{pmatrix}, \tag{8.4d}
$$

$$
\Theta_{d,12} = \begin{pmatrix} Q A_d^T & \hat{A}_{d,K}^T \\ A_d^T & A_d^T S^+ + C^T \hat{B}_{d,K}^T \end{pmatrix}, \tag{8.4e}
$$

$$
\Theta_{d,22} = -\begin{pmatrix} Q^+ & I \\ I & S^+ \end{pmatrix}, \tag{8.4f}
$$

$$
\Psi_{12} = I - Q\Gamma, \tag{8.4g}
$$

$$
\Psi_{13} = Q\Gamma^{1/2}, \tag{8.4h}
$$

$$
\Psi_{14} = N\Gamma_K^{1/2}, \tag{8.4i}
$$

$$
\Psi_{22} = S - \Gamma, \tag{8.4j}
$$

$$
\Delta_{11} = Q(t_0) R Q(t_0) + N(t_0) R_K N(t_0)^T. \tag{8.4k}
$$

$\square$

Proof The connection of systems (8.1a)–(8.1c) and (8.2a)–(8.2c) reads

$$
\dot{x}_{\mathrm{CL}}(t) = \begin{pmatrix} A_c(t) + B(t)D_K(t)C(t) & B(t)C_K(t) \\ B_K(t)C(t) & A_K(t) \end{pmatrix} x_{\mathrm{CL}}(t)
$$

$$
= A_{\mathrm{CL}}(t)x_{\mathrm{CL}}(t), \quad t > t_0, \quad t \notin \mathcal{T},
$$

---

[1] In order to avoid awkward notation, we discard, when possible, the time dependence both in (8.3a)–(8.3d) and (8.4a)–(8.4k).

$$x_{\mathrm{CL}}(t_k^+) = \begin{pmatrix} A_d(t_k) & 0 \\ B_{d,K}(t_k)C(t_k) & A_{d,K}(t_k) \end{pmatrix} x_{\mathrm{CL}}(t_k)$$

$$= A_{d,\mathrm{CL}}(t_k)x_{\mathrm{CL}}(t_k), \quad t_k \in \mathcal{T},$$

where $x_{\mathrm{CL}} = [x^T \ x_K^T]^T$. According to Theorem 7.2, it follows that Problem 8.1 is solvable *if and only if* there exists a piecewise continuous symmetric matrix-valued function $P(\cdot)$ such that

$$\dot{P}(t) + A_{\mathrm{CL}}(t)^T P(t) + P(t)A_{\mathrm{CL}}(t) < 0, \quad t \in [t_0, t_0 + T], \quad t \notin \mathcal{T}, \tag{8.5a}$$

$$\begin{pmatrix} -P(t_k) & A_{d,\mathrm{CL}}(t_k)^T P^+(t_k) \\ P^+(t_k)A_{d,\mathrm{CL}}(t_k) & -P^+(t_k) \end{pmatrix} < 0, \quad t_k \in \mathcal{T}, \tag{8.5b}$$

$$P(t) > \mathrm{diag}\big(\Gamma(t), \Gamma_K(t)\big), \quad t \in [t_0, t_0 + T], \tag{8.5c}$$

$$P(t_0) < \mathrm{diag}(R, R_K). \tag{8.5d}$$

We will prove the theorem by construction. According to Lemma 3.1, define the matrix-valued functions $P(\cdot)$, $\Pi_1(\cdot)$, and $\Pi_2(\cdot)$ satisfying (3.14a)–(3.14b), which in turn, by definition, satisfy (3.15a)–(3.15c). We now prove that, with the given choice of $P(\cdot)$, conditions (8.5a)–(8.5d) are equivalent to (8.3a)–(8.3d).

Indeed, by pre- and post-multiplying inequalities (8.5a), (8.5c), and (8.5d) by $\Pi_1(t)^T$ and $\Pi_1(t)$, respectively, conditions (8.3a), (8.3c), and (8.3d) follow once we define $\hat{A}_K(\cdot)$, $\hat{B}_K(\cdot)$, and $\hat{C}_K(\cdot)$ as in (3.16a)–(3.16d). Eventually, note that (8.3c) implies (3.10).

Moreover, by pre- and post-multiplying inequality (8.5b) by

$$\mathrm{diag}\big(\Pi_1(t_k)^T, \Pi_1^+(t_k)^T\big)$$

and

$$\mathrm{diag}\big(\Pi_1(t_k), \Pi_1^+(t_k)\big),$$

respectively, condition (8.3b) follows once we let

$$\hat{A}_{d,K}(t_k) = M^+(t_k)A_{d,K}(t_k)N(t_k)^T + M^+(t_k)B_{d,K}(t_k)C(t_k)Q(t_k)$$
$$+ S^+(t_k)A_d(t_k)Q(t_k), \tag{8.6a}$$

$$\hat{B}_{d,K}(t_k) = M^+(t_k)B_{d,K}(t_k). \tag{8.6b}$$

$\Diamond$

Similarly to what has been done in the case of CT-LTV systems in Chap. 3 (see Remark 3.2), the following procedure can be applied to compute the controller matrices.

Procedure 8.1 (Controller Design) If the hypotheses of Theorem 8.1 are fulfilled, in order to design the controller, the following steps have to be undertaken:

1. Find $Q(\cdot)$, $S(\cdot)$, $N(\cdot)$, $\hat{A}_K(\cdot)$, $\hat{B}_K(\cdot)$, $\hat{C}_K(\cdot)$, $D_K(\cdot)$, $\hat{A}_{d,K}(\cdot)$, and $\hat{B}_{d,K}(\cdot)$ such that conditions (8.3a)–(8.3d) are satisfied.
2. Calculate the matrix function $M(t) = (I - S(t)Q(t))N^{-T}(t)$ and its derivative $\dot{M}(t) = -(\dot{S}(t)Q(t) + S(t)\dot{Q}(t) + M(t)\dot{N}(t)^T)N(t)^{-T}$.
3. Obtain $A_K(\cdot)$, $B_K(\cdot)$, and $C_K(\cdot)$ by inverting (3.16a)–(3.16d), and $A_{d,K}(\cdot)$ and $B_{d,K}(\cdot)$ by inverting (8.6a)–(8.6b).   ◇

Concerning the nonlinear matrix inequality (8.3d), the comments done at the end of Sect. 3.3 still hold. Regarding the search for the nonsingular matrix $N(\cdot)$, required by the statement of Theorem 8.1, Lemma 3.2 is still valid.

From the previous theorem the following necessary and sufficient condition for finite-time stabilization of TD-IDLSs via state-feedback can be easily derived.

Corollary 8.1 *Problem 8.1 is solvable via state-feedback control if and only if there exist a piecewise continuously differentiable symmetric matrix-valued function $Q(\cdot)$ and a continuous matrix-valued function $L(\cdot)$ such that*

$$-\dot{Q}(t) + A_c(t)Q(t) + Q(t)A_c(t)^T + L(t)^T B(t)^T + B(t)L(t) < 0,$$

$$t \in [t_0, t_0 + T], \quad t \notin \mathcal{T}, \tag{8.7a}$$

$$\begin{pmatrix} -Q(t_k^+) & A_d(t_k)Q(t_k) \\ Q(t_k)A_d(t_k)^T & -Q(t_k) \end{pmatrix} < 0, \quad t \in \mathcal{T}, \tag{8.7b}$$

$$Q(t) < \Gamma^{-1}(t), \quad t \in [t_0, t_0 + T], \tag{8.7c}$$

$$Q(t_0) > R^{-1}. \tag{8.7d}$$

In this case, a controller gain that solves Problem 8.1 via state feedback is given by $K(t) = L(t)Q^{-1}(t)$.   □

Remark 8.1 As it has been discussed in the previous chapter (see Remarks 7.5 and 7.6), Theorem 8.1 and Corollary 8.1 also hold for TD-SLS since all the optimization variables in the corresponding feasibility problems are assumed to be piecewise continuous, with discontinuities in correspondence of the resetting times. The reservoir system example, considered in Sect. 8.5.2, will illustrate this fact. Furthermore, by a proper choice of the matrix-valued functions $\Gamma(\cdot)$ and $\Gamma_K(\cdot)$ the results for finite-time stabilization presented in this section also hold when the family (7.4) is composed of systems with different dimensions.   ◇

8.4 Finite-Time Stabilization of SD-IDLSs

In this section, we discuss how to extend the results given in Sect. 8.3 to the case of SD-IDLS.

Before deriving sufficient conditions to solve Problem 8.1 for SD-IDLSs, notice that for this class of IDLSs, the resetting times are not a priori known (see also the discussion in Sect. 7.5).

It should be remarked that for what concerns the controller design, the knowledge of the resetting set $\mathcal{D}$ cannot be exploited to obtain sufficient conditions in the form of LMIs. Indeed, differently from what it has been done in Sect. 7.5, by using the S-procedure, in [25] it has been shown that, even in the case of finite-time stabilization via state-feedback, the synthesis problem turns out to be a problem of feasibility of bilinear matrix inequalities [81].

It follows that, in order to recast the design problem in terms of LMIs, for SD-IDLSs, the controller design problem requires, at the price of some conservativeness, that condition (8.3b) be satisfied for all $t \in [t_0, t_0 + T]$ (not only at the resetting times).

The next remark briefly summarizes how to extend Theorem 8.1 and Corollary 8.1 to the case of SD-IDLSs.

Remark 8.2 (Feedback Control of SD-IDLS) For SD-IDLSs, Problem 8.1 is solvable:

- via output-feedback if there exist matrix-valued functions $Q(\cdot)$, $S(\cdot)$, $N(\cdot)$, $\hat{A}_K(\cdot)$, $\hat{B}_K(\cdot)$, $\hat{C}_K(\cdot)$, $D_K(\cdot)$, $\hat{A}_{d,K}(\cdot)$, and $\hat{B}_{d,K}(\cdot)$ such that conditions (8.3c) and (8.3d) hold, while both conditions (8.3a) and (8.3b) must be satisfied for all t in $[t_0, t_0 + T]$;
- via state-feedback if there exist two matrix-valued functions $Q(\cdot)$ and $L(\cdot)$ such that conditions (8.7c) and (8.7d) hold, while both conditions (8.7a) and (8.7b) must be satisfied for all t in $[t_0, t_0 + T]$.   ◇

8.5 Examples

This section presents two examples where the results for finite-time stabilization provided in this chapter are exploited. The first numerical example deals with the finite-time stabilization of an SD-IDLS. Then, we consider an engineering example that deals with three interconnected reservoirs. This last example shows how to apply the results presented in Sect. 8.3 to the case of TD-SLSs (see also Remark 8.1).

8.5.1 SD-IDLS

Consider the SD-IDLS

$$\dot{x}(t) = Ax(t) + Bu(t), \quad x(0) = x_0,$$
$$x^+(t) = A_d x(t), \quad x \in \mathcal{D},$$

Fig. 8.1 Finite-time stabilization of the SD-IDLS considered in Sect. 8.5.1. Temporal profiles of the two time-varying gains of the state-feedback controller

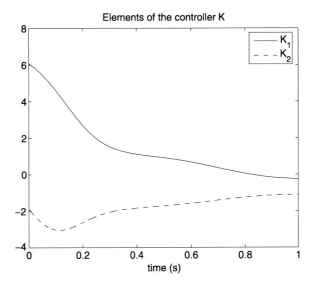

with

$$A = \begin{pmatrix} 0.5 & -1.5 \\ 0.5 & 0.5 \end{pmatrix}, \qquad B = \begin{pmatrix} 0 \\ 1 \end{pmatrix}, \qquad A_d = \begin{pmatrix} 0.8 & 0 \\ 0 & 0.8 \end{pmatrix},$$

and the resetting set

$$\mathcal{D} = \text{conv} \left(\begin{pmatrix} 0.5 \\ 0.2 \end{pmatrix}, \begin{pmatrix} 0.4 \\ 0.4 \end{pmatrix} \right).$$

For the considered SD-IDLS with $u = 0$, the conditions of Theorem 7.3 are not verified if we let $t_0 = 0$, $T = 1$ s, and

$$\Gamma = \begin{pmatrix} 1 & 0 \\ 0 & 1 \end{pmatrix}, \qquad R = \begin{pmatrix} 2.5 & 0 \\ 0 & 2.5 \end{pmatrix}.$$

We now propose to exploit Corollary 8.1 to design a state-feedback control $u(t) = K(t)x(t)$ that stabilizes in finite-time the considered SD-IDLS (see Remark 8.2). In order to do that, the matrix-valued functions $P(\cdot)$ and $L(\cdot)$ are assumed to be piecewise linear and are computed by means of the MATLAB LMI Toolbox. Given the solution of the D/DLMI feasibility problem, it is possible to design a state-feedback control law that makes the system finite-time stable wrt $(0, 1, R, \Gamma)$. The time evolution of the two components of the controller gain $K(t)$ is shown in Fig. 8.1.

Fig. 8.2 Schematic
representation of a connected
system of reservoirs
considered in Sect. 8.5.2

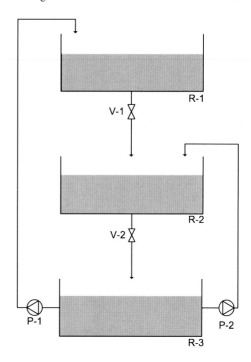

8.5.2 Reservoir System

In this section, the conditions of Corollary 8.1 are applied to design a state-feedback controller for a reservoir system modeled as a TD-SLS.

Figure 8.2 shows the system dealt with, made of three interconnected reservoirs R-1, R-2, and R-3. The two-state (ON/OFF) isolation valves V-1 and V-2, when open, let the liquid flow under the effect of gravity. We assume that these valves are alternatively open for given amounts of time and that they are activated by an external controller. The two pumps P-1 and P-2 move the liquid from R-3 to R-1 and R-2, respectively.

Moreover, the pump P-j, $j = 1, 2$, can be controlled only when V-j is open, whereas it is stopped otherwise. The overall objective is to control the amount of liquid transferred by the bottom reservoir R-3 toward the other two reservoirs using the pumps P-1 and P-2.

In this case the nonlinear system behavior can be approximated as an SLS as in (7.5a)–(7.5b), where the resetting law A_d is constant and equal to the identity matrix.

Indeed, considering a linear dependency of the outflow from the amount of liquid contained in the reservoirs, the system can be represented by

$$A_1 = \begin{pmatrix} -\frac{k_1}{a_1} & 0 & 0 \\ \frac{k_1}{a_2} & 0 & 0 \\ 0 & 0 & 0 \end{pmatrix}, \quad B_1 = \begin{pmatrix} \frac{1}{a_1} & 0 \\ 0 & 0 \\ -\frac{1}{a_3} & 0 \end{pmatrix}, \tag{8.8a}$$

Table 8.1 Model and parameters for the controller design used in the example in Sect. 8.5.2

Parameter	Value
a_i	1 m
k_j	10^{-3} m^2/s
T	60 s

Fig. 8.3 Time evolution of the weighting matrix $\Gamma(\cdot)$ considered for the example of connected reservoirs. Since $\Gamma(\cdot)$ is a diagonal matrix, only the elements on the diagonal are reported

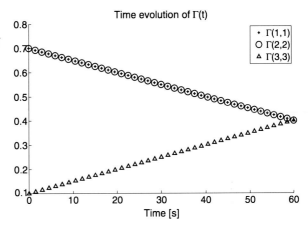

$$A_2 = \begin{pmatrix} 0 & 0 & 0 \\ 0 & -\frac{k_2}{a_2} & 0 \\ 0 & \frac{k_2}{a_3} & 0 \end{pmatrix}, \qquad B_2 = \begin{pmatrix} 0 & 0 \\ 0 & \frac{1}{a_2} \\ 0 & -\frac{1}{a_3} \end{pmatrix}, \qquad (8.8b)$$

where

- x_i is the height of the liquid in the ith reservoir, $i = 1, \ldots, 3$,
- a_i is the area of the ith reservoir, $i = 1, \ldots, 3$,
- k_j is the storage coefficient of the jth reservoir, $j = 1, \ldots 2$,

while u_j is the amount of liquid transferred by P-j from R-3 toward R-1, $j = 1, 2$.

It should be noticed that the model is valid as far as the states x_i remain nonnegative. This assumption is verified a posteriori, as it will be shown in the following. The values of the model parameters are summarized in Table 8.1.

Let us now suppose that the liquid is initially stored in R-3 and consider the goal of equally distributing it among the three reservoirs in 60 seconds. This problem can be casted in the finite-time stabilization framework by considering the weighting matrix $\Gamma(\cdot)$ shown in Fig. 8.3. Indeed, since $\Gamma(\cdot)$ is diagonal, each element on the main diagonal weights only one state. As shown in Fig. 8.3, the weight of the third state satisfies $\Gamma_{3,3}(0) = 0$; for $t > 0$, it increases until reaching the final value of 0.4 when $t = 60$ s. The weights of the other two states follow the opposite time evolution starting from the value of 0.7. Since the three weights converge toward the same value, the controller tries to distribute the liquid equally among the three reservoirs in order to minimize the state weighted norm. Figure 8.4 shows the switching signal $\sigma(t)$.

Fig. 8.4 Switching signal $\sigma(\cdot)$ for the example of connected reservoirs

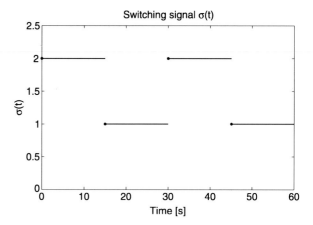

Fig. 8.5 Time trace of the state variables for the example of connected reservoirs. Note that the time evolution of the state variables is driven by the time evolution of the weighting matrix-valued function $\Gamma(\cdot)$ shown in Fig. 8.3

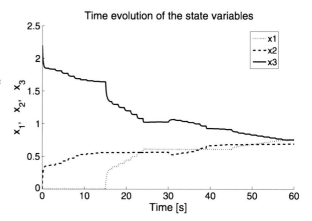

Taking into account Remark 8.1, it is possible to apply Corollary 8.1 in order to design a state-feedback control law that assures FTS wrt $(0, 60, R, \Gamma(\cdot))$ of the closed-loop system, where R is taken equal to the identity matrix. In particular, the optimization matrices needed to check the feasibility problem in Corollary 8.1 are assumed to be piecewise linear, with a different slope every 1.5 s.

Figures 8.5 and 8.6 show how the controller manages to obtain the expected result by keeping the state weighted norm always below 1. Due to the considered switching signal $\sigma(\cdot)$, the controller cannot act on the first reservoir during the first 15 seconds, and hence a big control action is taken right after the first switching instant, which implies a large reduction of the state weighted norm, as shown in Fig. 8.6.

8.6 Summary

In this chapter, the finite-time stabilization problem for IDLSs has been discussed. Starting from the analysis result based on the DLMI condition, stated in Theo-

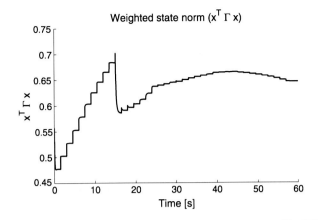

Fig. 8.6 Time trace of the state weighted norm for the example of connected reservoirs. Note that this norm has to be less than one in order to have finite-time stability for the closed-loop system

rem 7.2, necessary and sufficient conditions for finite-time stabilization of TD-IDLSs for both the state and output feedback cases have been derived.

Concerning SD-IDLSs, for the sake of brevity, the design procedure, based on Theorem 7.3, has been briefly detailed.

Then the application of the main results have been illustrated by two examples. In particular, it has been shown how the control of a real engineering plant, composed of three reservoirs, can be framed in the context of the FTS of TD-IDLSs. Such an example will be further investigated in the following chapter with the design of a controller guaranteeing robust FTS.

Chapter 9
Robustness Issues for IDLSs

9.1 Introduction

In this chapter, we consider IDLSs subject to various types of uncertainties.

First, we investigate the same framework of Chaps. 7 and 8, assuming, however, that the IDLS under consideration is affected by norm-bounded uncertainties. The mathematical machinery is, in this case, similar to that which allowed us to study the QFTS problem dealt with in Chap. 4.

Then, we focus ourselves on the class of SLSs introduced in Sect. 7.2.2, investigating the case in which the occurrence of the resetting times is unknown. This allows us to tackle real engineering situations, where the change of system dynamics is unpredictable and/or is due to an external triggering event.

The first result is a sufficient condition for FTS when the resetting times are known with a certain degree of uncertainty. In other words, we assume that each switching instant is known to belong to a given interval. Such a condition requires the solution of a suitable feasibility problem based on a coupled D/DLMI. Afterwards, we show that, as the uncertainty intervals reduce in size, such a condition becomes less conservative, becoming necessary and sufficient in the certain case (i.e., the resetting times are perfectly known).

Finally, the conceptually different situation in which the resetting times are totally unknown, namely the *arbitrary* switching case, is dealt with. Clearly, the sufficient condition for FTS obtained in this case is more restrictive (i.e., conservative) than those obtained in the uncertain switching case.

The analysis results are then used to derive sufficient conditions for the existence of state feedback controllers that finite-time stabilize the closed-loop system in the two cases above.

The jumping body and the reservoir system examples, introduced in Chaps. 7 and 8, are continued at the end of this chapter to illustrate the proposed techniques.

9.2 Uncertain IDLSs

In this section, we extend the approach considered in Chap. 4 to the class of IDLSs. In particular, norm-bounded uncertainties are added to the IDLS (7.1a)–(7.1b). First,

F. Amato et al., *Finite-Time Stability and Control*,
Lecture Notes in Control and Information Sciences 453,
DOI 10.1007/978-1-4471-5664-2_9, © Springer-Verlag London 2014

we consider uncertainties only for the continuous-time dynamics (7.1a); afterwards, norm-bounded uncertainties are also added to the resetting law (7.1b). Finally, the concepts of RFTS and QFTS are extended to the case of uncertain IDLSs.

As in Chap. 4, we model the uncertainty on the continuous-time dynamics as a linear fractional norm-bounded uncertainty, which leads to the definition of *uncertain* IDLS (U-IDLS)

$$\dot{x}(t) = \left(A_c(t) + F_c(t)\Delta_c(t)\left(I - H_c(t)\Delta_c(t)\right)^{-1}E_c(t)\right)x(t)$$

$$= \left(A_c(t) + F_c(t)\left(I - \Delta_c(t)H_c(t)\right)^{-1}\Delta_c(t)E_c(t)\right)x(t), \quad x(t_0) = x_0,$$

$$\left(t, x(t)\right) \notin \mathcal{S}, \tag{9.1a}$$

$$x^+(t) = A_d(t)x(t), \quad \left(t, x(t)\right) \in \mathcal{S}, \tag{9.1b}$$

where $F_c(\cdot)$, $E_c(\cdot)$, $H_c(\cdot)$, and $\Delta_c(\cdot)$ with $\|\Delta_c(t)\| \leq 1$, $t \in [0, +\infty[$, are matrix-valued functions of suitable dimensions. In particular, $\Delta_c(\cdot)$ is a normalized matrix-valued function that captures the uncertainty of the elements in $A_c(\cdot)$, while $F_c(\cdot)$, $E_c(\cdot)$, and $H_c(\cdot)$ take into account scaling factors.

Furthermore, supposing to add to (9.1a)–(9.1b) also a norm-bounded uncertainty on the resetting law, we obtain the following U-IDLS:

$$\dot{x}(t) = \left(A_c(t) + F_c(t)\Delta_c(t)\left(I - H_c(t)\Delta_c(t)\right)^{-1}E_c(t)\right)x(t), \quad x(t_0) = x_0,$$

$$\left(t, x(t)\right) \notin \mathcal{S}, \tag{9.2a}$$

$$x^+(t) = \left(A_d(t) + F_d(t)\Delta_d(t)\left(I - H_d(t)\Delta_d(t)\right)^{-1}E_d(t)\right)x(t), \quad \left(t, x(t)\right) \in \mathcal{S}, \tag{9.2b}$$

where, similarly to (9.1a)–(9.1b), $F_d(\cdot)$, $E_d(\cdot)$, and $H_d(\cdot)$ are matrix-valued function of suitable dimensions, and $\Delta_d(\cdot)$ is such that $\|\Delta_d(t)\| \leq 1$ for all t in $[0, +\infty[$.

Remark 9.1 Similarly to the case considered in Chap. 4, note that (9.1a), (9.2a), and (9.2b) are well posed if $I - H_*\Delta_*(t)$ is invertible for all $\Delta_*(t)$ with $\|\Delta_*(t)\| \leq 1$, where the subscript $*$ indicates either c or d.   ◇

The U-IDLS (9.2a)–(9.2b) can be equivalently rewritten as

$$\dot{x}(t) = \left(A_c(t) + \Delta A_c(t)\right)x(t), \quad x(t_0) = x_0, \quad \left(t, x(t)\right) \notin \mathcal{S}, \tag{9.3a}$$

$$x^+(t) = \left(A_d(t) + \Delta A_d(t)\right)x(t), \quad \left(t, x(t)\right) \in \mathcal{S}. \tag{9.3b}$$

We now extend the definitions of RFTS and QFTS given in Sect. 4.2 to the case of U-IDLSs (9.3a)–(9.3b).

Definition 9.1 (RFTS for U-IDLSs) Given an initial time t_0, a positive scalar T, a positive definite matrix R, and a positive definite matrix-valued function $\Gamma(\cdot)$, defined over $[t_0, t_0 + T]$, such that $\Gamma(t_0) < R$, system (9.3a)–(9.3b) is said to be robustly finite-time stable with respect to $(t_0, T, R, \Gamma(\cdot))$ if

$$x_0^T R x_0 \leq 1 \Rightarrow x(t)^T \Gamma(t) x(t) < 1, \quad t \in [t_0, t_0 + T],$$

for all $\Delta_c(\cdot)$, $\Delta_d(\cdot)$ with $\|\Delta_c(t)\|, \|\Delta_d(t)\| \leq 1, t \in [t_0, t_0 + T]$.   ◊

Definition 9.2 (QFTS for U-IDLSs)   Given an initial time t_0, a positive scalar T, a positive definite matrix R, and a positive definite matrix-valued function $\Gamma(\cdot)$, defined over $[t_0, t_0 + T]$, such that $\Gamma(t_0) < R$, system (9.3a)–(9.3b) is said to be quadratically finite-time stable with respect to $(t_0, T, R, \Gamma(\cdot))$ if and only if there exists a positive definite matrix-valued function $P(\cdot)$ that satisfies the following DLMI with terminal and initial conditions:

$$\dot{P}(t) + \left(A_c(t) + \Delta A_c(t)\right)^T P(t) + P(t)\left(A_c(t) + \Delta A_c(t)\right) < 0, \quad \left(t, x(t)\right) \notin \mathcal{S}, \tag{9.4a}$$

$$P(t) > \Gamma(t), \quad t \in [t_0, t_0 + T], \tag{9.4b}$$

$$P(t_0) < R, \tag{9.4c}$$

$$\left(A_d(t_i) + \Delta A_d(t_i)\right)^T P^+(t_i)\left(A_d(t_i) + \Delta A_d(t_i)\right) - P(t_i) < 0, \quad \left(t_i, x(t_i)\right) \in \mathcal{S}, \tag{9.4d}$$

for any admissible uncertainty of $\Delta_c(\cdot)$ and $\Delta_d(\cdot)$.   ◊

From the definitions given above it is easy to check that Lemma 4.1 holds also in the case of U-IDLSs.

9.2.1 QFTS of Time-Dependent U-IDLSs

Necessary and sufficient conditions for QFTS of time-dependent U-IDLSs are given in this section. The case of uncertain continuous dynamics (i.e., $\Delta A_d(\cdot) = 0$ for all $t \in [t_0, t_0 + T]$) is first considered; the proposed result is then extended to the more general case of U-IDLS given by (9.3a)–(9.3b).

Remark 9.2   For the sake of brevity, in this section, we do not deal with the finite-time stabilization problem. However, conditions to solve the finite-time stabilization problem by both state and output feedback can be easily derived by combining the results given in Sects. 9.2.1 and 9.2.2 with the approach presented in Sect. 4.3.   ◊

Theorem 9.1 (QFTS of Time-Dependent U-IDLSs; Uncertain Continuous-Time Dynamics)   *The time-dependent U-IDLSs (9.1a)–(9.1b) is QFTS wrt $(t_0, T, R, \Gamma(\cdot))$*

if and only if *there exist a piecewise continuously differentiable symmetric matrix-valued function* $P(\cdot)$ *and a function* $\lambda(\cdot) > 0$ *such that*

$$
\begin{pmatrix}
\dot{P}(t) + A_c^T(t)P(t) + P(t)A_c(t) + \lambda(t)E^T(t)_c E_c(t) & P(t)F_c(t) + \lambda E_c^T(t)H_c(t) \\
F_c^T(t)P(t) + \lambda H_c^T(t)E_c(t) & -\lambda(t)(I - H_c^T(t)H_c(t))
\end{pmatrix} < 0,
$$

$$
t \in [t_0, t_0 + T], \quad t \notin \mathcal{T}, \tag{9.5a}
$$

$$
A_d(t_k)^T P^+(t_k) A_d(t_k) - P(t_k) < 0, \quad t_k \in \mathcal{T}, \tag{9.5b}
$$

$$
P(t) > \Gamma(t), \quad t \in [t_0, t_0 + T], \tag{9.5c}
$$

$$
P(t_0) < R. \tag{9.5d}
$$

$\square$

Proof The proof easily follows exploiting the same arguments as in Theorem 4.1, which are based on the application of the *S*-procedure. $\square$

We now introduce the following lemma, taken from [42], which is exploited to extend Theorem 9.1 to the case of IDLSs with uncertainty on both the continuous dynamics and resetting law.

Lemma 9.1 [42] *Given two symmetric matrices* $\Pi_1, \Pi_2 \in \mathbb{R}^{n \times n}$, *a scalar* $\tau > 0$, *and four matrices* $A \in \mathbb{R}^{n \times n}$, $F \in \mathbb{R}^{n \times f}$, $E \in \mathbb{R}^{g \times n}$, *and* $H \in \mathbb{R}^{g \times f}$, *the condition*

$$
\begin{pmatrix}
-\Pi_1 & \Pi_1 A & \Pi_1 F & 0 \\
A^T \Pi_1 & -\Pi_2 & 0 & \tau E^T \\
F^T \Pi_1 & 0 & -\tau I & \tau H^T \\
0 & \tau E & \tau H & -\tau I
\end{pmatrix} < 0
$$

implies that for all $\Delta \in \mathbb{R}^{f \times g}$ *with* $\|\Delta\| \le 1$,

$$
\left(A + F\Delta(I - H\Delta)^{-1}E\right)^T \Pi_1\left(A + F\Delta(I - H\Delta)^{-1}E\right) - \Pi_2 < 0. \qquad \square
$$

It is now possible to prove the following result.

Theorem 9.2 (QFTS of Time-Dependent U-IDLSs; Uncertainties on Both the Continuous-Time Dynamics and the Resetting Law) *The time-dependent U-IDLSs* (9.3a)–(9.3b) *is QFTS wrt* $(t_0, T, R, \Gamma(\cdot))$ *if and only if there exist a piecewise continuously differentiable symmetric matrix-valued function* $P(\cdot)$ *and two func-*

tions $\lambda_1(\cdot), \lambda_2(\cdot) > 0$ *such that*

$$\begin{pmatrix} \dot{P}(t) + A_c^T(t)P(t) + P(t)A_c(t) + \lambda_1 E_c^T(t)E_c(t) & P(t)F_c(t) + \lambda_1 E_c^T(t)H_c(t) \\ F_c^T(t)P(t) + \lambda_1 H_c^T(t)E_c(t) & -\lambda_1(I - H_c^T(t)H_c(t)) \end{pmatrix} < 0,$$

$$t \in [t_0, t_0 + T], \quad t \notin \mathcal{T}, \tag{9.6a}$$

$$\begin{pmatrix} A_d^T(t_k)P^+(t_k)A_d(t_k) - P(t_k) + \lambda_2(t_k)E_d^T(t_k)E_d(t_k) & A_d^T(t_k)P^+(t_k)F_d(t_k) + \lambda_2(t_k)E_d^T(t_k)H_d(t_k) \\ F_d^T(t_k)P^+(t_k)A_d(t_k) + \lambda_2(t_k)H_d^T(t_k)E_d(t_k) & F_d^T(t_k)P^+(t_k)F_d(t_k) - \lambda_2(t_k)(I - H_d^T(t_k)H_d(t_k)) \end{pmatrix} < 0,$$

$$t_k \in \mathcal{T}, \tag{9.6b}$$

$$P(t) > \Gamma(t), \quad t \in [t_0, t_0 + T], \tag{9.6c}$$

$$P(t_0) < R. \tag{9.6d}$$

□

Proof In Theorem 9.1, inequality (9.6a) assures that $I - H_c\Delta_c(t)$ is nonsingular for all $\Delta_c(t)$ such that $\|\Delta_c(t)\| \leq 1$. By using similar arguments, inequality (9.6b) guarantees that a similar condition also holds for $I - H_d\Delta_d(t_k)$.

Next, we show that inequality (9.6b) implies[1]

$$\left(A_d + F_d\Delta_d(I - H_d\Delta_d)^{-1}E_d\right)^T P^+ \left(A_d + F_d\Delta_d(I - H_d\Delta_d)^{-1}E_d\right) - P < 0 \tag{9.7}$$

for all $t_k \in \mathcal{T}$; this allows us to prove the theorem by using similar arguments as in Theorem 9.1.

Indeed, by exploiting the result of Lemma 9.1 we get that if

$$\begin{pmatrix} -P^+ & P^+A_d & P^+F_d & 0 \\ A_d^T P^+ & -P & 0 & \lambda_2 E_d^T \\ F_d^T P^+ & 0 & -\lambda_2 I & \lambda_2 H_d^T \\ 0 & \lambda_2 E_d & \lambda_2 H_d & -\lambda_2 I \end{pmatrix} < 0, \quad t_k \in \mathcal{T}, \tag{9.8}$$

then condition (9.7) holds.

---

[1] In the following, the time argument is dropped in order to simplify the notation.

Applying Schur complements to (9.8), we obtain

$$
\begin{pmatrix}
-P^+ & P^+ A_d & P^+ F_d \\
A_d^T P^+ & -P + \lambda_2 E_d^T E_d & \lambda_2 E_d^T H_d \\
F_d^T P^+ & \lambda_2 H_d^T E_d & -\lambda_2 (I - H_d^T H_d)
\end{pmatrix} < 0, \quad t_k \in \mathcal{T}. \tag{9.9}
$$

Condition (9.6b) follows from (9.9) by applying Schur complements again. It follows that condition (9.6b) implies (9.7) since it is equivalent to (9.8). □

9.2.2 QFTS of State-Dependent U-IDLSs

This section provides a sufficient condition to check QFTS of state-dependent U-IDLSs. As already discussed in Chap. 7, in the case of state-dependent IDLSs, either certain or uncertain, the resetting times are not a priori known, which leads to sufficient conditions.

In particular, the following theorem deals with the more general case of U-IDLSs given by (9.3a)–(9.3b).

Theorem 9.3 (QFTS of State-Dependent U-IDLSs) *The state-dependent U-IDLSs (9.3a)–(9.3b) is QFTS wrt $(t_0, T, R, \Gamma(\cdot))$ if there exist a piecewise continuously differentiable symmetric matrix-valued function $P(\cdot)$ and two functions $\lambda_1(\cdot)$, $\lambda_2(\cdot) > 0$ such that*

$$
\left(
\begin{matrix}
\dot{P}(t) + A_c^T(t) P(t) + P(t) A_c(t) + \lambda_1(t) E_c^T(t) E_c(t) \\
F_c^T(t) P(t) + \lambda_1(t) H_c^T(t) E_c(t)
\end{matrix}
\right.
$$

$$
\left.
\begin{matrix}
P(t) F_c(t) + \lambda_1(t) E_c^T(t) H_c(t) \\
-\lambda_1(t)(I - H_c^T(t) H_c(t))
\end{matrix}
\right) < 0,
$$

$$
t \in [t_0, t_0 + T], \tag{9.10a}
$$

$$
x^T \left(
\begin{matrix}
A_d^T(t_k) P^+(t_k) A_d(t_k) - P(t_k) + \lambda_2(t_k) E_d^T(t_k) E_d(t_k) \\
F_d^T(t_k) P^+(t_k) A_d(t_k) + \lambda_2(t_k) H_d^T(t_k) E_d(t_k)
\end{matrix}
\right.
$$

$$
\left.
\begin{matrix}
A_d^T(t_k) P^+(t_k) F_d(t_k) + \lambda_2(t_k) E_d^T(t_k) H_d(t_k) \\
F_d^T(t_k) P^+(t_k) F_d(t_k) - \lambda_2(t_k)(I - H_d^T(t_k) H_d(t_k))
\end{matrix}
\right) x < 0,
$$

$$
t \in [t_0, t_0 + T], \quad x \in \mathcal{D}, \tag{9.10b}
$$

$$
P(t) > \Gamma(t), \quad t \in [t_0, t_0 + T], \tag{9.10c}
$$

$$
P(t_0) < R. \tag{9.10d}
$$

□

Proof The proof follows similar arguments to those used in the proof of Theorems 9.2 and 7.3.   □

It is worth noticing that condition (9.10b) can be turned into LMIs by using the *S*-procedure, as described in Sect. 7.5.2.

9.3 SLSs with Uncertain Resetting Times

This section deals with the SLSs introduced in Sect. 7.2.2 when the case of uncertain resetting times is considered. In particular, for this class of uncertain hybrid systems, we refer to the *classic* definition of FTS given in Sect. 7.3.

Concerning the knowledge of the resetting set $\mathcal{T}$, in this paper, we shall consider two cases.

Arbitrary Switching (AS): no information about the resetting times is available, i.e., the resetting times are totally unknown.

Uncertain Switching (US): the resetting times are known with a given uncertainty. Without loss of generality, we will assume that $t_j \in [\bar{t}_j - \Delta T_j, \bar{t}_j + \Delta T_j]$, where $\bar{t}_j$ is the *nominal* value of the jth resetting time t_j, $j = 1, \dots, h$.

Since the switching signal $\sigma(\cdot)$ is piecewise constant with discontinuities in correspondence of the resetting times, in the US case, it is useful to introduce the following notation:

$$\psi_1 =]t_0, \bar{t}_1 + \Delta T_1[,$$

$$\psi_j =]\bar{t}_{j-1} - \Delta T_{j-1}, \bar{t}_j + \Delta T_j[, \quad j = 2, \dots, h,$$

$$\psi_{h+1} =]\bar{t}_h - \Delta T_h, t_0 + T],$$

$$\phi_j = [\bar{t}_j - \Delta T_j, \bar{t}_j + \Delta T_j], \quad j = 1, \dots, h.$$

Given the above definitions, we make the following assumption.

Assumption 9.1 *The time intervals* ϕ_j, $j = 1, \dots, h$, *satisfy*

$$\bigcap_{j=1}^{h} \phi_j = \emptyset, \tag{9.11}$$

which implies the knowledge of the order of resetting times.   ◇

The following theorem introduces a sufficient condition to check FTS when the jth resetting time is known with a given uncertainty $\pm \Delta T_j$, i.e., when the US case is considered.

Theorem 9.4 (FTS in US Case, [24]) *Suppose that there exist $h + 1$ piecewise continuously differentiable symmetric matrix-valued functions $P_j(\cdot)$, $j = 1,\ldots, h + 1$, that satisfy the following D/DLMI:*

$$\dot{P}_j(t) + A_{\sigma(t_{j-1})}^T(t)P_j(t) + P_j(t)A_{\sigma(t_{j-1})}(t) < 0, \quad t \in \psi_j, j = 1,\ldots, h+1, \tag{9.12a}$$

$$J^T(t)P_{j+1}(t)J(t) - P_j(t) < 0, \quad t \in \phi_j, \quad j = 1,\ldots, h, \tag{9.12b}$$

$$P_j(t) > \Gamma(t), \quad t \in \psi_j, \quad j = 1,\ldots, h+1, \tag{9.12c}$$

$$P_1(t_0) < R. \tag{9.12d}$$

Then the SLS (7.5a)–(7.5b) is finite-time stable wrt $(t_0, T, R, \Gamma(\cdot))$ under the US assumption. ▫

Proof Let us start by noticing that, exploiting the knowledge of the order of resetting times implied by (9.11) (see Assumption 9.1), it is possible to *assign* a single optimization matrix $P_j(\cdot)$ and the corresponding active linear system enabled by $\sigma(t)$ to each time interval ψ_j, $j = 1,\ldots, h + 1$.

Suppose now that $\Delta T_j = 0$ for $j = 1,\ldots, h$; it turns out that the intervals ϕ_j become equal to $\{\bar{t}_j\}$ and

$$\bigcap_{j=1}^{h+1} \psi_j = \emptyset.$$

Hence, it is possible to define the following piecewise differentiable positive definite matrix-valued function $P(\cdot)$:

$$P(t) = P_j(t) \quad \text{for } t \in \psi_j.$$

Let

$$V(t, x) = x^T(t)P(t)x(t);$$

given a time instant $t \notin \mathcal{T}$, the derivative with respect to time reads

$$\frac{d}{dt}\left(x^T(t)P(t)x(t)\right) = x^T(t)\left(\dot{P}(t) + A_\sigma^T(t)P(t) + P(t)A_\sigma(t)\right)x(t),$$

which is negative definite in view of (9.12a). At the discontinuity point $\bar{t}_j$, we have

$$V\left(t_j^+, x\right) - V(t_j, x) = x^T\left(t_j^+\right)P\left(t_j^+\right)x\left(t_j^+\right) - x^T(t_j)P(t_j)x(t_j)$$

$$= x^T(t_j)\left(J^T(t_j)P\left(t_j^+\right)J(t_j) - P(t_j)\right)x(t_j),$$

which is negative definite by virtue of (9.12b). We can conclude that $V(t, x)$ is strictly decreasing along the trajectories of system (7.5a)–(7.5b). Hence, given x_0

such that $x_0^T R x_0 \leq 1$, we have

$$x^T(t)\Gamma(t)x(t) < x^T(t)P(t)x(t)$$

$$< x_0^T P(t_0)x_0$$

$$< x_0^T R x_0 \leq 1,$$

where the first inequality is guaranteed by (9.12c), and the third one by (9.12d).

Now consider the case of $\Delta T_j \neq 0$. First, notice that, although unknown, a resetting time set exists. Furthermore, knowing the number of the switching times and their order, we also know the time interval in which every switch can occur. It turns out that the previous proof still applies when considering the time interval ϕ_j in place of the time instant $\bar{t}_j$.

Indeed, conditions (9.12a) and (9.12c) still have to be verified in ψ_j, i.e., in the time interval in which the corresponding linear system is potentially active. Moreover, in the uncertain case, condition (9.12b) has to be checked in ϕ_j, i.e., in the time interval in which the state jump could occur. $\diamond$ $\square$

Remark 9.3 The conditions in Theorem 9.4 become less severe as the uncertainty intervals reduce in size. In the limit case of no uncertainty, the conditions stated in Theorem 9.4 recover the one stated in Theorem 7.2 (see also Remarks 7.5 and 7.6) and hence become also necessary for FTS. $\diamond$

The next corollary extends the result given in Theorem 9.4 to the AS case.

Corollary 9.1 (FTS in AS Case) *Suppose that there exist l piecewise continuously differentiable symmetric matrix-valued functions $P_i(\cdot)$, $i = 1, \ldots, l$, that satisfy the following D/DLMI:*

$$\dot{P}_i(t) + A_i^T(t)P_i(t) + P_i(t)A_i(t) < 0, \quad t \in \,]t_0, t_0 + T], \quad i \in \mathcal{P}, \qquad (9.13a)$$

$$J^T(t)P_i(t)J(t) - P_j(t) < 0, \quad t \in [t_0, t_0 + T], \quad i, j \in \mathcal{P}, \qquad (9.13b)$$

$$P_i(t) > \Gamma(t), \quad t \in [t_0, t_0 + T], \quad i \in \mathcal{P}, \qquad (9.13c)$$

$$P_i(t_0) < R, \quad i \in \mathcal{P}. \qquad (9.13d)$$

Then the SLS (7.5a)–(7.5b) is finite-time stable wrt $(t_0, T, R, \Gamma(\cdot))$ under the AS assumption. $\square$

Proof The case of AS can be seen as a special case of US, in which the uncertainty covers the considered time interval $[t_0, t_0 + T]$. If this is the case, the intervals ψ_j and ϕ_j, $j \in \mathcal{P}$, coincide with $[t_0, t_0 + T]$. As a consequence, for all i in $\mathcal{P}$, conditions (9.12a) and (9.12c) have to be verified in $[t_0, t_0 + T]$, while (9.12d) must be verified at t_0. It turns out that (9.13a), (9.13c), and (9.13d) hold.

Furthermore, condition (9.12b) has to be verified for all $i, j \in \mathcal{P}$ since, at each time instant, the system can switch between any linear dynamics defined in the family (7.4). Hence, (9.13b) holds, and since the resetting times are not a priori known, $P_j(\cdot)$, $j \in \mathcal{P}$, need to be differentiable.   ◊

The AS case introduces the maximum level of conservatism. In particular, since condition (9.13b) has to be verified for all $i, j \in \mathcal{P}$, $J(\cdot)$ needs to be Schur stable for all t in $[t_0, t_0 + T]$. In other words, due to the lack in the knowledge of resetting times, it is necessary to have *stable* resetting laws in order to check the FTS of SLS. Furthermore, each linear system in the family (7.4) has to be finite-time stable in order to render finite-time stable the SLS with uncertain resetting times.

Starting from the analysis conditions provided in Theorem 9.4, we now address the problem of finite-time stabilization of SLSs with uncertain resetting times. Only the US case is presented since the results for the AS case can be derived from the next theorem exploiting similar arguments as for Corollary 9.1.

Theorem 9.5 (Finite-Time Stabilization in the US Case, [24]) *In the case of uncertain switchings, Problem 8.1 is solvable via state-feedback control if there exist $h + 1$ piecewise continuously differentiable symmetric matrix-valued functions $\Pi_j(\cdot)$ and $h + 1$ matrix-valued functions $L_j(\cdot)$, $j = 1, \ldots, h + 1$, such that*

$$-\dot{\Pi}_j(t) + A_{\sigma(t_{j-1})}(t)\Pi_j(t) + \Pi_j(t)A_{\sigma(t_{j-1})}^T(t) + L_j^T(t)B_{\sigma(t_j)}^T(t) + B_{\sigma(t_j)}(t)L_j(t)$$

$$< 0, \quad t \in \psi_j, j = 1, \ldots, h + 1, \tag{9.14a}$$

$$\begin{pmatrix} -\Pi_{j+1}(t) & J(t)\Pi_j(t) \\ \Pi_j(t)J^T(t) & -\Pi_j(t) \end{pmatrix} < 0, \quad t \in \phi_j, j = 1, \ldots, h, \tag{9.14b}$$

$$\Pi_j(t) < \Gamma^{-1}(t), \quad t \in \psi_j, j = 1, \ldots, h + 1, \tag{9.14c}$$

$$\Pi_1(t_0) > R^{-1}. \tag{9.14d}$$

In this case, a controller gain that solves Problem 8.1 via state-feedback is given by

$$K(t) = \begin{cases} L_j(t)\Pi_j(t)^{-1}, & t \in [t_{j-1}, t_j), \quad j = 1, \ldots, h, \\ L_{h+1}(t)\Pi_{h+1}(t)^{-1}, & t \in [t_h, t_0 + T], \end{cases}$$

where $t_k, k = 1, \ldots, h$, is the kth resetting time.   □

Proof The proof readily follows applying Theorem 9.4 to the closed-loop system, letting $\Pi_j(t) = P_j^{-1}(t)$ and applying the Schur complements to (9.12b) (see also Corollary 8.1).   ◊

9.4 Examples

In this section, we recall the two examples considered in Sects. 7.7.3 and 8.5.2, and we apply the results introduced in this chapter.

9.4.1 Jumping Body

In this example, we add a norm-bounded uncertainty to the jumping body TD-IDLS considered in Sect. 7.7.3.

In particular, we considered the following equation system:

$$\dot{x}_1(t) = x_2(t),$$
$$\dot{x}_2(t) = -\frac{c}{m}x_2(t),$$
$$\dot{x}_3(t) = x_4(t), \qquad \text{for } t \neq \frac{i}{f}, i = 1, 2, \ldots,$$
$$\dot{x}_4(t) = -\frac{c}{m}x_4(t),$$
$$\dot{x}_5(t) = 0,$$

$$x_1^+(t) = x_1(t),$$
$$x_2^+(t) = x_2(t) + \zeta\Delta_{11}x_2(t) + \zeta\Delta_{12}x_4(t) + \zeta\Delta_{13}x_5(t),$$
$$x_3^+(t) = x_3(t), \qquad \text{for } t_i = \frac{i}{f}, i = 1, 2, \ldots,$$
$$x_4^+(t) = x_4(t) + \zeta\Delta_{21}x_2(t) + \zeta\Delta_{22}x_4(t) + \zeta\Delta_{23}x_5(t),$$
$$x_5^+(t) = x_5(t),$$

where Δ_{ij}, $i, j = 1, 2, 3$, are norm-bounded uncertainties.

The considered uncertain IDLS can be written as in (9.2a)–(9.2b) with

$$A_c = \begin{pmatrix} 0 & 1 & 0 & 0 & 0 \\ 0 & -\frac{c}{m} & 0 & 0 & 0 \\ 0 & 0 & 0 & 1 & 0 \\ 0 & 0 & 0 & -\frac{c}{m} & 0 \\ 0 & 0 & 0 & 0 & 0 \end{pmatrix}, \quad A_d = \begin{pmatrix} 1 & 0 & 0 & 0 & 0 \\ 0 & 1 & 0 & 0 & 0 \\ 0 & 0 & 1 & 0 & 0 \\ 0 & 0 & 0 & 1 & 0 \\ 0 & 0 & 0 & 0 & 1 \end{pmatrix},$$

$$\Delta = \begin{pmatrix} \Delta_{11} & \Delta_{12} & \Delta_{13} \\ \Delta_{21} & \Delta_{22} & \Delta_{23} \end{pmatrix}, \quad E_d = \begin{pmatrix} 0 & 1 & 0 & 0 & 0 \\ 0 & 0 & 0 & 1 & 0 \\ 0 & 0 & 0 & 0 & 1 \end{pmatrix},$$

$$F_d = \zeta \begin{pmatrix} 0 & 0 \\ 1 & 0 \\ 0 & 0 \\ 0 & 1 \\ 0 & 0 \end{pmatrix}, \quad H_d = \begin{pmatrix} 0 & 0 \\ 0 & 0 \\ 0 & 0 \end{pmatrix},$$

without any uncertainty on the continuous-time dynamics.

We assume that the elastic surface shown in Fig. 7.7 has a radius $r_f = 7$ m and that the body jumps on it starting from a circular subregion of radius $r_i = 5$ m, with a jump frequency $f = 0.5$ Hz and $\zeta = 0.2$. Similarly to what has been done in Sect. 7.7.3, we exploit Theorem 9.2 to find the maximum $\mathcal{N} \in \mathbb{N}$ such that the body does not exit the elastic surface during the time interval $[0, \mathcal{N}]$, taking into account the model uncertainty. Choosing the R and Γ matrices as in Sect. 7.7.3 and solving the D/DLMI feasibility problem (9.6a)–(9.6d) by means of an optimization tool, it turns out that the maximum value of $\mathcal{N}$ is equal to 28.

9.4.2 Reservoirs System

In this example, we consider the reservoir system modeled as TD-SLS introduced in Sect. 8.5.2, and we assume that the switching signal $\sigma(\cdot)$ shown in Fig. 8.4 is affected by a time uncertainty of $\Delta T = 3$ s.

Given this uncertainty on the resetting times, we want to design a robust state-feedback control law that assures the reservoir system to be finite-time stable wrt $(0, 60, I, \Gamma(\cdot))$, where the matrix-valued function $\Gamma(\cdot)$ is the one reported in Fig. 8.3.

It turns out that, given $\Delta T = 3$ s, the feasibility problem in Theorem 9.5 admits a solution, and hence it is possible to design the desired control law.

9.5 Summary

In this chapter, the FTS problem for U-IDLSs has been discussed. More precisely, two different kinds of uncertainties have been considered: norm-bounded model uncertainties of the same form considered in Chap. 4 and uncertainties in the resetting law.

For IDLSs whose continuous-time parts are affected by uncertainties in norm-bounded form, a necessary and sufficient condition for QFTS has been provided in the TD case; such a condition is then extended to the case in which the uncertainty also enters the resetting law dynamics. When the resetting law is state dependent, the condition turns out to be only sufficient due to the inherent conservativeness introduced by the circumstance that the resetting times are not a priori known. It is worth noting that Theorems 9.1–9.3 are stated in this book for the first time.

The second part of the chapter is essentially based on the recent paper [24]; only TD-IDLSs (or, more generally, time-dependent SLSs) are considered, and the case where the system dynamics are certain but the switching instants are not a priori known is considered. Two different cases are discussed; in the first case (AS), it is assumed that no information about the resetting times is available, that is, the resetting times are unknown; in the second case (US), it is assumed that the resetting times are known with a given uncertainty.

In both cases, sufficient conditions in terms of optimization problems involving DLMIs are provided; obviously, the condition in the AS case is much more severe than the condition derived under the US assumption.

The examples introduced in Chap. 7 (jumping body system) and in Chap. 8 (reservoir system) are used to illustrate the benefits of the proposed technique.

This area of research is currently rather active among control researchers, with a special attention to the so-called dwell-time approach; see, for example, [41, 46, 68, 83].

References

1. Amato, F.: Robust Control of Linear Systems Subject to Uncertain and Time-Varying Parameters. Springer, Berlin (2006)
2. Amato, F., Ariola, M.: Finite-time control of discrete-time linear systems. IEEE Trans. Autom. Control **50**, 724–729 (2005)
3. Amato, F., Pironti, A., Scala, S.: Necessary and sufficient conditions for quadratic stability and stabilizability of uncertain linear time-varying systems. IEEE Trans. Autom. Control **41**(1), 125–128 (1996)
4. Amato, F., Ariola, M., Dorato, P.: Robust finite-time stabilization of linear systems depending on parametric uncertainties. In: Proc. IEEE Conference on Decision and Control, pp. 1207–1208 (1998)
5. Amato, F., Ariola, M., Abdallah, C.T., Dorato, P.: Finite-time control for uncertain linear systems with disturbance inputs. In: Proc. American Control Conference, San Diego, CA, pp. 1776–1780 (1999)
6. Amato, F., Ariola, M., Dorato, P.: Finite time control of linear systems subject to parametric uncertainties and disturbances. Automatica **37**, 1459–1463 (2001)
7. Amato, F., Ariola, M., Cosentino, C.: Finite-time control with pole placement. In: Proc. European Control Conference, Cambridge, UK (2003)
8. Amato, F., Ariola, M., Cosentino, C., Abdallah, C.T., Dorato, P.: Necessary and sufficient conditions for finite-time stability of linear systems. In: Proc. American Control Conference, Denver, CO (2003)
9. Amato, F., Carbone, M., Ariola, M., Cosentino, C.: Finite-time stability of discrete-time systems. In: Proc. American Control Conference, Boston, MA, pp. 1440–1444 (2004)
10. Amato, F., Ariola, M., Cosentino, C.: Finite-time control of linear time-varying systems via output feedback. In: Proc. American Control Conference, Portland, OR, pp. 4722–4726 (2005)
11. Amato, F., Ariola, M., Carbone, M., Cosentino, C.: Finite-time output feedback control of linear systems via differential linear matrix conditions. In: Proc. IEEE Conf. on Decision and Control, San Diego, CA, pp. 5371–5375 (2006)
12. Amato, F., Ariola, M., Cosentino, C.: Finite-time stabilization via dynamic output feedback. Automatica **42**, 337–342 (2006)
13. Amato, F., Ambrosino, R., Ariola, M., Calabrese, F.: Finite-time stability of linear systems: an approach based on polyhedral Lyapunov functions. In: Proc. IEEE Conf. on Decision and Control, New Orleans, LO, pp. 1100–1105 (2007)
14. Amato, F., Ambrosino, R., Ariola, M., Calabrese, F.: Finite-time stability analysis of linear discrete-time systems via polyhedral Lyapunov functions. In: Proc. American Control Conference, Seattle, WA, pp. 1656–1660 (2008)

F. Amato et al., *Finite-Time Stability and Control*,
Lecture Notes in Control and Information Sciences 453,
DOI 10.1007/978-1-4471-5664-2, © Springer-Verlag London 2014

15. Amato, F., Ambrosino, R., Ariola, M., Cosentino, C.: Finite-time stability of linear time-varying systems with jumps. Automatica **45**(5), 1354–1358 (2009)

16. Amato, F., Ambrosino, R., Ariola, M., Calabrese, F.: Finite-time stability of linear systems: an approach based on polyhedral Lyapunov functions. IET Control Theory Appl. **4**, 167–1774 (2010)

17. Amato, F., Ambrosino, R., Cosentino, C., De Tommasi, G.: Input-output finite-time stabilization of linear systems. Automatica **46**, 1558–1562 (2010)

18. Amato, F., Ariola, M., Cosentino, C.: Finite-time control of discrete-time linear systems: analysis and design conditions. Automatica **46**, 919–924 (2010)

19. Amato, F., Ariola, M., Cosentino, C.: Finite-time stability of linear-time-varying systems: analysis and controller design. IEEE Trans. Autom. Control **55**, 1003–1008 (2010)

20. Amato, F., Cosentino, C., Merola, A.: Sufficient conditions for finite-time stability and stabilization of nonlinear quadratic systems. IEEE Trans. Autom. Control **55**(2), 430–434 (2010)

21. Amato, F., Ambrosino, R., Ariola, M., De Tommasi, G.: Robust finite-time stability of impulsive dynamical linear systems subject to norm-bounded uncertainties. Int. J. Robust Nonlinear Control **21**(10), 1080–1092 (2011)

22. Amato, F., Ambrosino, R., Cosentino, C., De Tommasi, G.: Finite-time stabilization of impulsive dynamical linear systems. Nonlinear Anal. Hybrid Syst. **5**(1), 89–101 (2011)

23. Amato, F., Ariola, M., Cosentino, C.: Robust finite-time stabilisation of uncertain linear systems. Int. J. Control **84**(12), 2117–2127 (2011)

24. Amato, F., Carannante, G., De Tommasi, G.: Finite-time stabilization of switching linear systems with uncertain resetting times. In: Proc. Mediterranean Control Conference, Corfu, Greece, pp. 1361–1366 (2011)

25. Amato, F., Carannante, G., De Tommasi, G.: Input-output finite-time stabilisation of a class of hybrid systems via static output feedback. Int. J. Control **84**(6), 1055–1066 (2011)

26. Amato, F., Carannante, G., De Tommasi, G., Pironti, A.: Input-output finite-time stability of linear systems: necessary and sufficient conditions. IEEE Trans. Autom. Control **57**(12), 3051–3063 (2012)

27. Amato, F., De Tommasi, G., Pironti, A.: Necessary and sufficient conditions for finite-time stability of impulsive dynamical linear systems. Automatica **49**, 2546–2550 (2013)

28. Ambrosino, R., Calabrese, F., Cosentino, C., De Tommasi, G.: Sufficient conditions for finite-time stability of impulsive dynamical systems. IEEE Trans. Autom. Control **54**(4), 861–865 (2009)

29. Ambrosino, R., Garone, E., Ariola, M., Amato, F.: Piecewise quadratic functions for finite-time stability analysis. In: Proc. IEEE Conf. on Decision and Control, Florence, Italy, pp. 6535–6540 (2012)

30. Anderson, B.D.O., Moore, J.B.: Optimal Control: Linear Quadratic Methods. Prentice Hall, Upper Saddle River (1989)

31. Barmish, B.R.: Stabilization of uncertain systems via linear control. IEEE Trans. Autom. Control **28**(8), 848–850 (1983)

32. Basar, T., Bernhard, P.: H_∞-Optimal Control and Related Minimax Design Problems: A Dynamic Game Approach. Birkhäuser, Boston (1991)

33. Becker, G., Packard, A.: Robust performance of linear parametrically varying systems using parametrically-dependent linear feedback. Syst. Control Lett. **23**, 205–215 (1994)

34. Bhat, S.P., Bernstein, D.S.: Finite-time stability of continuous autonomous systems. SIAM J. Control Optim. **38**(3), 751–766 (2000)

35. Blanchini, F.: Nonquadratic Lyapunov functions for robust control. Automatica **31**(3), 451–461 (1995)

36. Boel, R., Stremersch, G.: Hybrid Systems II. Springer, Berlin (1995)

37. Boel, R., Stremersch, G.: Discrete Event Systems: Analysis and Control. Springer, Berlin (2000)

38. Boyd, S., El Ghaoui, L., Feron, E., Balakrishnan, V.: Linear Matrix Inequalities in System and Control Theory. SIAM, Philadelphia (1994)

39. Brayton, R.K., Tong, C.H.: Constructive stability and asymptotic stability of dynamical systems. IEEE Trans. Circuits Syst. **27**(11), 1121–1130 (1980)
40. Chen, H., Guo, K.-H.: Constrained $\mathcal{H}_\infty$ control of active suspensions: an LMI approach. IEEE Trans. Control Syst. Technol. **13**, 412–421 (2005)
41. Chen, G., Yang, Y.: Finite-time stability of switched positive linear systems. Int. J. Robust Nonlinear Control (2012). doi:10.1002/rnc.2870
42. Chilali, M., Gahinet, P., Apkarian, P.: Robust pole placement in LMI regions. IEEE Trans. Autom. Control **44**, 2257–2270 (1999)
43. Corless, M.: Robust stability analysis and controller design with quadratic Lyapunov functions. In: Variable Structure and Lyapunov Control. Lectures Notes in Control and Information Sciences, vol. 193, pp. 181–203. Springer, Berlin (1994)
44. Dorato, P.: Short time stability in linear time-varying systems. In: Proc. IRE Int. Convention Record Pt. 4, pp. 83–87 (1961)
45. Doyle, J.C., Packard, A., Zhou, K.: Review of LFTs, LMIs and μ. In: Proc. IEEE Conference on Decision and Control, Brighton, UK, pp. 1227–1232 (1991)
46. Du, H., Lin, X., Li, S.: Finite-time boundedness and stabilization of switched linear systems. Kibernetika **46**, 870–889 (2010)
47. Gahinet, P.: Explicit controller formulas for LMI-based H_∞ synthesis. Automatica **32**, 1007–1014 (1996)
48. Gahinet, P., Nemirovski, A., Laub, A.J., Chilali, M.: LMI Control Toolbox. The Mathworks, Natick (1995)
49. Garcia, G., Tarbouriech, S., Bernussou, J.: Finite-time stabilization of linear time-varying continuous systems. IEEE Trans. Autom. Control **54**, 364–369 (2009)
50. Gayek, J.E.: A survey of techniques for approximating reachable and controllable sets. In: Proc. IEEE Conf. on Decision and Control, Brighton, UK, pp. 1724–1729 (1991)
51. Geromel, J.C., Peres, P.L.D., Bernussou, J.: On a convex parameter space method for linear control design of uncertain systems. SIAM J. Control Optim. **29**, 381–402 (1991)
52. Grantham, W.J.: Estimating controllability boundaries for uncertain systems. In: Vincent, T.L., Skowronski, J.M. (eds.) Renewable Resource Management, pp. 151–162 (1980)
53. Grantham, W.J.: Estimating reachable sets. J. Dyn. Syst. Meas. Control **103**(4), 420–422 (1981)
54. Haddad, W.M., Chellaboina, V., Nersesov, S.G.: Impulsive and Hybrid Dynamical Systems. Princeton University Press, Princeton (2006)
55. Hahn, H.: Stability of Motion. Springer, Berlin (1967)
56. Hong, Y., Huang, J., Xu, Y.: On an output feedback finite-time stabilization problem. IEEE Trans. Autom. Control **46**(2), 305–308 (2001)
57. Hong, Y., Jiang, Z.P., Feng, G.: Finite-time input-to-state stability and applications to finite-time control design. SIAM J. Control Optim. **48**(7), 4395–4418 (2010)
58. Hu, T., Blanchini, F.: Non-conservative matrix inequality conditions for stability/stabilizability of linear differential inclusions. Automatica **46**, 190–196 (2010)
59. Jakubovič, V.A.: The S-procedure in linear control theory. Vestn. Leningr. Univ., Math. **4**, 73–93 (1977)
60. Johansson, M., Rantzer, A.: Computation of piecewise quadratic Lyapunov functions for hybrid systems. IEEE Trans. Autom. Control **43**(4), 555–559 (1998)
61. Johansson, K.H., Egerstedt, M., Lygeros, J., Sastry, S.S.: On the regularization of Zeno hybrid automata. Syst. Control Lett. **38**, 141–150 (1999)
62. Kamenkov, G.: On stability of motion over a finite interval of time. J. Appl. Math. Mech. **17**, 529–540 (1953) (in Russian)
63. Khalil, H.K.: Nonlinear Systems. Prentice Hall, Upper Saddle River (2002)
64. Lebedev, A.: On stability of motion during a given interval of time. J. Appl. Math. Mech. **18**, 139–148 (1954) (in Russian)
65. Lebedev, A.: The problem of stability in a finite interval of time. J. Appl. Math. Mech. **18**, 75–94 (1954) (in Russian)
66. Liberzon, D.: Impulsive Control Theory. Springer, Berlin (2001)

67. Liberzon, D.: Switching in Systems and Control. Springer, Berlin (2003)
68. Liu, H., Shen, Y.: H_∞ finite-time control for switched linear systems with time-varying delay. Intell. Control Autom. **2**, 203–213 (2011)
69. Liu, L., Sun, J.: Finite-time stabilization of linear systems via impulsive control. Int. J. Control **81**, 905–909 (2008)
70. Lofberg, J.: YALMIP: a toolbox for modeling and optimization in Matlab. In: Proc. IEEE Symposium on Computer-Aided Control System Design, Taipei, Taiwan, pp. 284–289 (2004)
71. Mastellone, S., Dorato, P., Abdallah, C.T.: Finite-time stability of discrete-time nonlinear systems: analysis and design. In: Proc. IEEE Conference on Decision and Control, Paradise Island, Bahamas (2004)
72. Michel, A., Porter, D.: Practical stability and finite-time stability of discontinuous systems. IEEE Trans. Circuit Theory **CT-19**, 123–129 (1972)
73. Moulay, E., Perruquetti, W.: Finite time stability conditions for non-autonomous continuous systems. Int. J. Control **81**(5), 797–803 (2008)
74. Nersesov, S.G., Haddad, W.M.: Finite-time stabilization of nonlinear impulsive dynamical systems. Nonlinear Anal. Hybrid Syst. **2**, 832–845 (2008)
75. Packard, A., Balas, G., Safonov, M., Chiang, R., Gahinet, P., Nemirovski, A.: Robust Control Toolbox. The Mathworks, Natick (1984–2006)
76. Pettersson, S.: Analysis and Design of Hybrid Systems. Ph.D. Thesis, Chalmers University of Technology (1999)
77. Shaked, U., Suplin, V.: A new bounded real lemma representation for the continuous-time case. IEEE Trans. Autom. Control **46**(9), 1420–1426 (2001)
78. Shen, Y.: Finite-time control of linear parameter-varying systems with norm-bounded exogenous disturbance. J. Control Theory Appl. **6**(2), 184–188 (2008)
79. Sun, Z.: Stability of piecewise linear systems revisited. Annu. Rev. Control **34**(2), 221–231 (2010)
80. Sun, J., Xu, J., Yue, D.: Stochastic finite-time stability of nonlinear Markovian switching systems with impulsive effects. J. Dyn. Syst. Meas. Control **134**, 011011 (2011). doi:10.1115/1.4005359
81. Van Antwerp, J., Braatz, R.: A tutorial on linear and bilinear matrix inequalities. J. Process Control **10**(4), 363–385 (2000)
82. Wang, Y., Shi, X., Wang, G., Zuo, Z.: Finite-time stability for continuous-time switched systems in the presence of impulse effects. IET Control Theory Appl. **6**, 1741–1744 (2012)
83. Wang, Y., Wang, G., Shi, X., Zuo, Z.: Finite-time stability analysis of impulsive switched discrete-time linear systems: the average dwell time approach. Circuits Syst. Signal Process. **31**, 1877–1886 (2012)
84. Wang, Y., Shi, X., Zuo, Z., Chen, M.Z.Q., Shao, Y.: On finite-time stability for nonlinear impulsive switched systems. Nonlinear Anal., Real World Appl. **14**, 807–814 (2013)
85. Weiss, L., Infante, E.F.: Finite time stability under perturbing forces and on product spaces. IEEE Trans. Autom. Control **12**, 54–59 (1967)
86. Xu, J., Sun, J.: Finite-time stability of linear time-varying singular impulsive systems. IET Control Theory Appl. **4**, 2239–2244 (2010)
87. Xu, J., Sun, J.: Finite-time stability of nonlinear switched impulsive systems. Int. J. Syst. Sci. **44**, 889–895 (2013)
88. Yang, Y., Li, J., Chen, G.: Finite-time stability and stabilization of nonlinear stochastic hybrid systems. J. Math. Anal. Appl. **356**, 338–345 (2009)
89. Zhao, S., Sun, J., Liu, L.: Finite-time stability of linear time-varying singular systems with impulsive effects. Int. J. Control **81**(11), 1824–1829 (2008)
90. Zhou, K., Doyle, J.C., Glover, K.: Robust and Optimal Control. Prentice Hall, Upper Saddle River (1996)
91. Zuo, Z., Liu, Y., Wang, Y., Li, H.: Finite-time stochastic stability and stabilisation of linear discrete-time Markovian jump systems with partly unknown transition probabilities. IET Control Theory Appl. **6**, 1522–1526 (2012)

Index

F. Amato et al., *Finite-Time Stability and Control*,
Lecture Notes in Control and Information Sciences 453,
DOI 10.1007/978-1-4471-5664-2, © Springer-Verlag London 2014

Printed by Publishers' Graphics LLC
FMRO140502.23.34.8